PHYSICAL CHEMISTRY RESEARCH FOR ENGINEERING AND APPLIED SCIENCES

VOLUME 3

High Performance Materials and Methods

PHYSICAL CHEMISTRY RESEARCH FOR ENGINEERING AND APPLIED SCIENCES

VOLUME 3

High Performance Materials and Methods

Edited by

Eli M. Pearce, PhD, Bob A. Howell, PhD,
Richard A. Pethrick, PhD, DSc, and
Gennady E. Zaikov, DSc

Apple Academic Press Inc. | Apple Academic Press Inc.
3333 Mistwell Crescent | 9 Spinnaker Way
Oakville, ON L6L 0A2 | Waretown, NJ 08758
Canada | USA

©2015 by Apple Academic Press, Inc.

First issued in paperback 2021

Exclusive worldwide distribution by CRC Press, a member of Taylor & Francis Group

No claim to original U.S. Government works

ISBN 13: 978-1-77463-094-5 (pbk)
ISBN 13: 978-1-77188-058-9 (hbk)

Library and Archives Canada Cataloguing in Publication

Physical chemistry research for engineering and applied sciences/edited by Eli M. Pearce, PhD, Bob A. Howell, PhD, Richard A. Pethrick, PhD, DSc, and Gennady E. Zaikov, DSc.

Includes bibliographical references and index.
Contents: Volume 3. High performance materials and methods.
ISBN 978-1-77188-058-9 (v. 3 : bound)
1. Chemistry, Physical and theoretical. 2. Chemistry, Technical.
3. Physical biochemistry. I. Pearce, Eli M., author, editor II. Howell, B. A. (Bobby Avery), 1942-, author, editor III. Pethrick, R. A. (Richard Arthur), 1942-, author, editor IV. Zaikov, G. E. (Gennadii Efremovich), 1935-, author, editor

QD453.3.P49 2015   541   C2015-900409-8

Library of Congress Cataloging-in-Publication Data

Physical chemistry research for engineering and applied sciences/Eli M. Pearce, PhD, Bob A. Howell, PhD, Richard A. Pethrick, PhD, DSc, and Gennady E. Zaikov, DSc.
volumes cm
Includes bibliographical references and index.
Contents: volume 1. Principles and technological implications -- volume 2. Polymeric materials and processing -- volume 3. High performance materials and methods
ISBN 978-1-77188-053-4 (alk. paper)
1. Chemistry, Physical and theoretical. 2. Chemistry, Technical. 3. Physical biochemistry. I. Pearce, Eli M. II. Howell, B. A. (Bobby Avery), 1942- III. Pethrick, R. A. (Richard Arthur), 1942- IV. Zaikov, G. E. (Gennadii Efremovich), 1935-

QD453.3.P49 2015   541--dc23   2015000878

Apple Academic Press also publishes its books in a variety of electronic formats. Some content that appears in print may not be available in electronic format. For information about Apple Academic Press products, visit our website at **www.appleacademicpress.com** and the CRC Press website at **www.crcpress.com**

ABOUT THE EDITORS

Eli M. Pearce, PhD

Dr. Eli M. Pearce was the President of the American Chemical Society. He served as Dean of the Faculty of Science and Art at Brooklyn Polytechnic University in New York, as well as a Professor of Chemistry and Chemical Engineering. He was the Director of the Polymer Research Institute, also in Brooklyn. At present, he consults for the Polymer Research Institute. As a prolific author and researcher, he edited the *Journal of Polymer Science* (Chemistry Edition) for 25 years and was an active member of many professional organizations.

Bob A. Howell, PhD

Bob A. Howell, PhD, is a Professor in the Department of Chemistry at Central Michigan University in Mount Pleasant, Michigan. He received his PhD in physical organic chemistry from Ohio University in 1971. His research interests include flame-retardants for polymeric materials, new polymeric fuel-cell membranes, polymerization techniques, thermal methods of analysis, polymer-supported organoplatinum antitumor agents, barrier plastic packaging, bioplastics, and polymers from renewable sources.

Richard A. Pethrick, PhD, DSc

Professor R. A. Pethrick, PhD, DSc, is currently a Research Professor and Professor Emeritus in the Department of Pure and Applied Chemistry at the University of Strathclyde, Glasglow, Scotland. He was Burmah Professor in Physical Chemistry and has been a member of the staff there since 1969. He has published over 400 papers and edited and written several books. Recently, he has edited several publications concerned with the techniques for the characterization of the molar mass of polymers and also the study of their morphology. He currently holds a number of EPSRC grants and is involved with Knowledge Transfer Programmes involving three local companies involved in production of articles made out of polymeric materials. His current research involves AWE and has acted as a consultant for BAE Systems in the area of explosives and a company involved in the production of anticorrosive coatings.

Dr. Pethrick is on the editorial boards of several polymer and adhesion journals and was on the Royal Society of Chemistry Education Board. He is a Fellow of the Royal Society of Edinburgh, the Royal Society of Chemistry, and the Institute of Materials, Metal and Mining. Previously, he chaired the 'Review of Science Provision 16-19' in Scotland and the restructuring of the HND provision in chemistry. He was also involved in the creation of the revised regulations for accreditation by the Royal Society of Chemistry of the MSc level qualifications in chemistry. For a many years, he was the Deputy Chair of the EPSRC IGDS panel and involved in a number of reviews of the courses developed and offered under this program. He has been a member of the review panel for polymer science in Denmark and Sweden and the National Science Foundation in the USA.

Gennady E. Zaikov, DSc

Gennady E. Zaikov, DSc, is the Head of the Polymer Division at the N. M. Emanuel Institute of Biochemical Physics, Russian Academy of Sciences, Moscow, Russia, and Professor at Moscow State Academy of Fine Chemical Technology, Russia, as well as Professor at Kazan National Research Technological University, Kazan, Russia.

He is also a prolific author, researcher, and lecturer. He has received several awards for his work, including the Russian Federation Scholarship for Outstanding Scientists. He has been a member of many professional organizations and on the editorial boards of many international science journals.

Physical Chemistry Research for Engineering and Applied Sciences in 3 Volumes

Physical Chemistry Research for Engineering and Applied Sciences:
Volume 1: Principles and Technological Implications
Editors: Eli M. Pearce, PhD, Bob A. Howell, PhD,
Richard A. Pethrick, PhD, DSc, and Gennady E. Zaikov, DSc

Physical Chemistry Research for Engineering and Applied Sciences:
Volume 2: Polymeric Materials and Processing
Editors: Eli M. Pearce, PhD, Bob A. Howell, PhD,
Richard A. Pethrick, DSc, PhD, and Gennady E. Zaikov, DSc

Physical Chemistry Research for Engineering and Applied Sciences:
Volume 3: High Performance Materials and Methods
Editors: Eli M. Pearce, PhD, Bob A. Howell, PhD,
Richard A. Pethrick, DSc, PhD, and Gennady E. Zaikov, DSc

CONTENTS

LIST OF CONTRIBUTORS

M. V. Bazunova
Scientific Degree the Candidate of the Chemical Sciences, Post the Docent of the Department of High-Molecular Connections and General Chemical Technology of the Chemistry Faculty of the Bashkir State University, Official Address: 450076, Ufa, Zaks Validi Street, 32, Tel. (Official): (347) 229-96-86, Mobile: 89276388192, Bashkir State University, 32 Zaki Validi Street, 450076 Ufa, Republic of Bashkortostan, Russia, E-mail: mbazunova@mail.ru

Yu. V. Berestneva
Donetsk National University, 24 Universitetskaya Street, 83 055 Donetsk, Ukraine

Sanjay Kumar Bharti
School of Pharmaceutical Sciences, Guru Ghasidas Vishwa Vidyalaya (A Central University), Bilaspur-495009, Chattisgarh, India

V. I. Binyukov
The Federal State Budget Institution of Science N.M. Emanuel Institute of Biochemical Physics, Russian Academy of Sciences, 4 Kosygin str., Moscow, 119334 Russia

Valentina Chernova
Bashkir State University, Ufa, Russia, 450074, Zaki Validi st., 32, E-mail: onlyalena@mail.ru

M. R. El-Aassar
Polymer materials research Department, Institute of Advanced Technology and New Material, City of Scientific Research and Technology Applications, New Borg El-Arab City 21934, Alexandria, Egypt

M. S. Mohy Eldin
Polymer materials research Department, Institute of Advanced Technology and New Material, City of Scientific Research and Technology Applications, New Borg El-Arab City 21934, Alexandria, Egypt

Alfiya Galina
Bashkir State University, Ufa, Russia, 450074, Zaki Validi st., 32, E-mail: onlyalena@mail.ru

Y. G. Galyametdinov
Doctor of chemical sciences, head of department of Physical and Colloid Chemistry, KNRTU, E-mail: office@kstu.ru.

L. Ya. Gendel
Biological Faculty, Moscow State University, Moscow, 119899 Russia, Emanuel Institute of Biochemical Physics, Russian Academy of Sciences, Moscow, 117 977, Russia

E. A. Hassan
Department of Chemistry, Faculty of Science, Al-Azhar University, Cairo/Egypt

S. V. Kolesov
Scientific Degree the Doctor of the Chemical Sciences, Post the Professor of the Department of High-Molecular Connections and General Chemical Technology of the Chemistry Faculty of the Bashkir State University, Official address: 450076, Ufa, Zaks Validi street, 32, Tel. (Official): (347) 229-96-86, Institute of Biochemical Physics named N. M. Emanuel of Russian Academy of Sciences, 4 Kosygina Street, 119334, Moscow, Russia, E-mail: kolesovservic@mail.ru

N. N. Komova
Moscow State University of Fine Chemical Technology, 86 Vernadskii prospekt, Moscow 119571, Russia, E-mail: Komova_@mail.ru

Elena I. Kulish
Bashkir State University, Ufa, Russia, 450074, Zaki Validi st., 32, E-mail: alenakulish@rambler.ru

Roman Lazdin
Bashkir State University, Ufa, Russia, 450074, Zaki Validi st., 32, E-mail: alenakulish@rambler.ru

A. M. Lipanov
Institute of Mechanics, Ural Branch of the Russian Academy of Sciences, T. Baramsinoy 34, Izhevsk, Russia E-mail: postmaster@ntm.udm.ru

Debarshi Kar Mahapatra
School of Pharmaceutical Sciences, Guru Ghasidas Vishwa Vidyalaya (A Central University), Bilaspur-495009, India, Tel: +91 7552-260027, Fax: +91 7752-260154; E-mail: mahapatradebarshi@gmail.com

L. I. Matienko
The Federal State Budget Institution of Science N. M. Emanuel Institute of Biochemical Physics, Russian Academy of Sciences, 4 Kosygin str., Moscow, 119334 Russia, E-mail: matienko@sky.chph.ras.ru

E. M. Mil
The Federal State Budget Institution of Science N.M. Emanuel Institute of Biochemical Physics, Russian Academy of Sciences, 4 Kosygin str., Moscow, 119334 Russia

L. A. Mosolova
The Federal State Budget Institution of Science N.M. Emanuel Institute of Biochemical Physics, Russian Academy of Sciences, 4 Kosygin str., Moscow, 119334, Russia

A. I. Nigmatullina
Chemistry and Processing Technology of Elastomers Department, Kazan National Research Technological University, 68 K. Marks str., Kazan, Russia

A. A. Nikiforov
Chemistry and Processing Technology of Elastomers Department, Kazan National Research Technological University, 68 K. Marks str., Kazan, Russia

N. A. Okhotina
Chemistry and Processing Technology of Elastomers Department, Kazan National Research Technological University, 68 K. Marks str., Kazan, Russia

V. V. Osipova
Docent of Department of Physical and Colloid Chemistry, KNRTU

O. A. Panfilova
Chemistry and Processing Technology of Elastomers Department, Kazan National Research Technological University, 68 K. Marks str., Kazan, Russia

E. Yu. Parshina
Biological Faculty, Moscow State University, Moscow, 119899 Russia, Emanuel Institute of Biochemical Physics, Russian Academy of Sciences, Moscow, 117 977 Russia

Valiev Denis Radikovich

Scientific degree: Post the Student of the Department of High-Molecular Connections and General Chemical Technology of the Chemistry Faculty of the Bashkir State University, Official address: 450076, Ufa, Zaks Validi street, 32. Tel. (Official): (347) 229-96-86, E-mail: valief@mail.ru

E. V. Raksha
Donetsk National University, 24 Universitetskaya Street, 83 055 Donetsk, Ukraine

A. B. Rubin
Biological Faculty, Moscow State University, Moscow, 119899 Russia, Emanuel Institute of Biochemical Physics, Russian Academy of Sciences, Moscow, 117 977 Russia

Angela Shurshina
Bashkir State University, Ufa, Russia, 450074, Zaki Validi st., 32, E-mail: alenakulish@rambler.ru

R. F. Tukhvatullin
Bashkir State University, 32 Zaki Validi Street, 450076 Ufa, Republic of Bashkortostan, Russia

Irina Tuktarova
Bashkir State University, Ufa, Russia, 450074, Zaki Validi st., 32, E-mail: onlyalena@mail.ru

N. A. Turovskij
Donetsk National University, 24 Universitetskaya Street, 83 055 Donetsk, Ukraine, E-mail: N.Turovskij@donnu.edu.ua

A. V. Vakhrushev
Institute of Mechanics, Ural Branch of the Russian Academy of Sciences, T. Baramsinoy 34, Izhevsk, Russia E-mail: postmaster@ntm.udm.ru

S. I. Volfson
Chemistry and Processing Technology of Elastomers Department, Kazan National Research Technological University, 68 K. Marks str., Kazan, Russia

G. E. Zaikov
Institute of Biochemical Physics named N.M. Emanuel of Russian Academy of Sciences, 4 Kosygina Street, 119334, Moscow, Russia, Official address: 4 Kosygina Street, 119334, Moscow, Russia, E-mail: chembio@sky.chph.ras.ru

L. O. Zaskokina
Master of department of Physical and Colloid Chemistry, KNRTU

M. Yu. Zubritskij
Institute of Biochemical Physics, Russian Academy of Sciences, 4 Kosygin Street, 117 334 Moscow, Russian Federation

LIST OF ABBREVIATIONS

ADRs	Adverse Drug Reactions
AFM	Atomic-Force Microscopy
AIREBO	Adaptive Intermolecular Reactive Empirical Bond Order
AMA	American Medical Association
ARD	Acireductone Dioxygenase
BD	Brownian Dynamics
Bu	Butyl Radical
CAP	College of American Pathologists
CNT	Carbon Nanotube
CNV	Copy Number Variants
CO	Carbon Monoxide
CS	Cellulose
CTAB	Cetyl Trimethyl Ammonium Bromide
De	Diffusion coefficient
DFT	Density Function Method
DMF	Di Methyl Form amide
DNMT	DNA Methyl Transferases
DPD	Dissipative Particle Dynamics
DSC	Differential Scanning Calorimetry
DTC	Direct-To-Consumer
DWNTs	Double Walled Carbon Nanotubes
ER	Estrogen Receptor
FDA	Food and Drug Administration
FEM	Finite Element Method
GCK	Glucokinase
GINA	Genetic Information Non-discrimination Act
GSDBT	Generalized Shear Deformation Beam Theory
GWAS	Genome-Wide Association Studies
HGP	Human Genome Project
His	Histidine-Ligands
HIT	Healthcare Information Technology
HMPA	Hexa Methyl Phosphorotri Amide
HNF-1α	Hepatocyte Nuclear Factor-1α
HNF-1β	Hepatocyte Nuclear Factor-1β
HO	Haem Oxygenase

HP	Hydro Peroxide
HTS	High Temperature Shearing
IPS	Intellectual Property System
LB	Lattice Boltzmann
LBHBs	Low-Barrier Hydrogen Bonds
LD	Local Density
LDPE	Low Density Polyethylene
mAbs	Monoclonal Antibodies
MC	Monte Carlo
MD	Molecular Dynamics
MiR	MircoRNA
MMT	Montmorillonite
MODY	Maturity-Onset Diabetes of Young
MSP	Methionine Salvage Pathway
MWNTs	Multi-Walled Nano Tubes
NC	Nano Carbon
NIH	National Institutes of Health
NO	Nitrogen Monoxide
OCT	Organic Cation Transporter
PBQ	p-benzoquinone
PE	Polyethylene
PEH	Phenyl Ethyl Hydroperoxide
PhTEOS	Phenyl Trimethoxysilane
PM	Personalized Medicine
PP	Polypropylene
PPARG	Peroxisome Proliferator-Activated Receptor-γ
PSDT	Parabolic Shear Deformation Theory
PTO	Patent and Trademark Office
PZT	Plumbum Zirconate Titanate
QC	Quasi-Continuum
QM	Quantum Mechanics
RBM	Radial Breathing Mode
RVE	Representative Volume Element
SJS	Stevense-Johnson syndrome
SNPs	Single Nucleotide Polymorphisms
SubSIP	Substrate Separated Ion Pair
SWNTs	Single Walled Nano Tubes
TBMD	Tight Bonding Molecular Dynamics
TEN	Toxic Epidermal Necrolysis
TG	Thermogravimetric Method
TSGs	Tumor Suppressor Genes
UC	University of Cincinnati

UV-spectrum	Ultra Violet-Spectrum
VIM	Variational Iteration Method
VKORC1	Vitamin K Epoxide Reductase Complex Subunit 1
VMOS	Vinyl Trimethoxysilane

LIST OF SYMBOLS

A	adsorption capacity
A_{hi}	hydrodynamic invariant
$A_0, A_1,...,A_5$	functions of Eshelby's tensor
B	initial expansion rate
[B]	matrix containing the derivation of the shape function
C	concentration of carbon black in polypropylene
c	concentration of the polymer in the solution (g/dL)
[D]	elasticity matrix
d	density of the adsorbate
E	electrophiles
E_f	fiber modulus
E_m	modulus of the matrix
$\vec{F}_{bi}$	vector direction
$\vec{F}_i(t)$	random set of forces at a given temperature
$\vec{F}_i(t)$	force acting on the i-th atom or particle at time t
$G_0 = 2e^2/h$	transmission of single ballistic quantum channel
$G_s(t)$	concentration of the desorbed substance at time t
H (i) and H (j)	Hamiltonian
H_{melt}	enthalpy of polymer fusion phase
k	rate constant
K_B	Boltzmann constant
K_C	*equilibrium* constant of the complex formation (dm³·mol⁻¹)
k_d	rate constant of the complex decomposition (s⁻¹)
k_o	characteristic parameter for pure PP
k_1	constant reflecting the reaction of the polymer with a solvent
k2	rate constant for chain propagation
k5	rate constant for quadratic chain termination
m_i	atomic mass
m_0	weight of solvent in pycnometer
N_k	mass of the i-th atom
n	kinetic parameter
N_A	Avogadro's number
N_k	number of atoms forming each nanoparticle

$\vec{\rho}_{cj}$	vector connecting points c and j
$\vec{\rho}_{ij}$	radius-vector determining the position of the i-th atom relative to the j-th atom
$\vec{r}_1$	atomic position
R_m^n and r_m^n	undeformed and deformed lengths of truss elements
$\tilde{S}$	averaged selectivity of ethyl benzene oxidation to PEH
S_0	sedimentation constant
S_{lim}	arbitrary value
$\{T\}$	surface traction vector
T_c	crystallization temperature
t	time
U	internal energy
V	volume
v_f	fiber volume fraction
$\vec{V}_{i0}, \vec{V}_i$	initial and current velocities of the i-th atom
$\vec{x}_{i0}, \vec{x}_i$	original and current coordinates of the i-th atom
w_i	measure of the metal activity
w_{PEH}	rate of PEH accumulation

GREEK SYMBOLS

α	proportion of reacted material
α_i	friction coefficient in the atomic structure
ΔG	total free energy of activation
ΔG_η	free enthalpy of activation of molecular diffusion across the phase boundary
$\Delta\delta_{max}$	difference between the chemical shift of -CO-OH group proton of complex-bonded and free hydro peroxide ($\Delta\delta_{max} = \delta_{comp} - \delta_{ROOH}$)
ΔU	change in the sum of the mixing energy
ΔH_{melt0}	melting enthalpy of the polymer with 100% crystallinity
Ω	integration is over the entire domain
Ω_k	area occupied by the nano element
Φ_{cb}	chemical bonds
Φ_{es}	electrostatics
Φ_{pg}	flat groups
Φ_{hb}	hydrogen bonds

$\Phi(\bar{\rho}_{ij})$	potential depending on the mutual positions of all the atoms
Φ_{ta}	torsion angles
Φ_{va}	valence angles
Φ_{vv}	Vander Waals contacts
$\bar{\sigma}_{fi}$	fibers average stress
$\bar{\sigma}_{i}$	composite average stress
$\bar{\sigma}_{mi}$	matrix average stress,
$\{\mathfrak{I}\}$	force vector
τ_0	characteristic catalyst lifetime
$[\eta]$	intrinsic viscosity, dL/g
$\mathfrak{y}_{sp}$	specific viscosity of polymer solution
η_0	dynamic viscosity of solvent,
ρ_T	electrical resistivity at temperature T
ρ_0	density of the solvent g/cm^3
ρ_{20}	resistivity at room temperature
v	partial specific volume, cm^3/g
v_0	volume of the pycnometer
υ_0	Poisson's ratio of the matrix, parameters
ζ	shape parameter

PREFACE

High performance materials and their composites have attracted great attention in recent years due to their unique combination of toughness, lightness and dimensional stability. High performance materials and their composites have captured the attention of scientists and engineers alike as materials that offer outstanding properties for high performance applications.

Over the last decade, several new families of high performance materials and their composites have been reported which find enhanced application potential in the more challenging industrial areas. Such materials and their composites provide an improved set of properties like higher service temperatures at extreme conditions and good mechanical strength, dimensional stability, thermal degradation resistance, environmental stability, gas barrier, solvent resistance, and electrical properties.

This volume presents the various categories of high performance materials and their composites and provides up-to-date synthesis details, properties, characterization, and applications for such systems in order to give readers and users better information to select the required material.

The volume provides the following features:

- inclusion of a wide range of high performance and engineering materials;
- details on the synthesis and properties of each of new materials;
- orientation towards the practical industrial applications;
- chapters from some of the world's most well-known and respected experts in the field.

CHAPTER 1

REGULARITY OF OXIDATION OF WASTE FIBROUS AND FILM MATERIALS OF POLYETHYLENE AND POLYPROPYLENE: A RESEARCH NOTE

M. V. BAZUNOVA, S. V. KOLESOV, R. F. TUKHVATULLIN, E. I. KULISH, and G. E. ZAIKOV

CONTENTS

ABSTRACTS

Physicochemical laws of surface oxidation of materials of polypropylene, polyethylene and their waste are studied. Found that the background polyolefin material has virtually no effect on the course of its oxidative modification.

1.1   INTRODUCTION

The purpose of chemical modification of polymers is to change chemical structure by introducing the functional groups with different chemical nature into the macromolecules. In some cases, it is necessary to improve the characteristics of a polymer surface by chemical modification while retaining the properties of materials in the volume and shape of the polymer material (fiber, film, bulk product). It is necessary, for example, when changing the wettability, sorption, adhesion and electrical characteristics of materials in the desired direction [1].

An effective method of modifying of the surface of polymeric material is oxidation. It is known that at temperatures below 90 °C the reaction takes place mainly on the surface and is not accompanied by oxidative degradation of the polymer in the bulk. Obvious, that it is impractical to create complex chemical surface modification technology simultaneously saving money on technical costs of creating the necessary forms of materials. Oxidation is also a convenient method of preactivation of the polymer surface, which leads to the appearance of oxygen-containing functional groups capable of being active centers during the further chemical modification. For example, it is known that the polyolefins oxidation is accompanied by the formation of hydro peroxide (HP) groups (Fig. 1.1). Further thermal decomposition of the HP groups leads to the appearance of free radicals on the surface and initiate growth of the grafted chains.

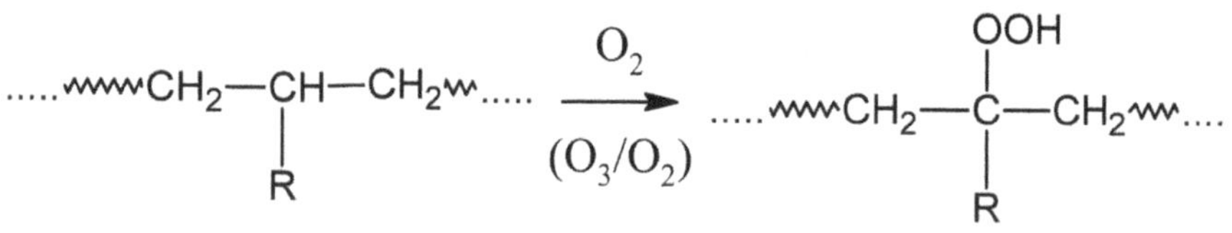

FIGURE 1.1   Oxidation of carbon-chain polymers (R = $-CH_3$, $-H$).

Therefore, it seems reasonable to study the simple and technologically advanced methods of surface oxidation of the waste fibrous and film materials of polypropylene (PP) and polyethylene (PE) as a method of creating of the secondary polymeric materials.

1.2 EXPERIMENTAL PART

Previously cleaned and dried waste PP-fibers (original PP-fiber GOST 26574–85), waste PE-film (original PE-film GOST 17811–17878) have been used as an object of research in this paper.

Concentration of HP-groups on the surface of oxidized samples is defined by modified iodometric method [2] with the photoelectric calorimeter КФК-2МП.

The limited wetting angle is defined by the standard procedure [3].

The adsorption capacity (A) of the samples under static conditions for condensed water vapor, determined by method of complete saturation of the sorbent by adsorbate vapor in standard conditions at 20 °C [4] and calculated by the formula: $A=m/(M \times d)$, where in m – mass of the adsorbed water, g; M – mass of the dried sample, g; d – density of the adsorbate, g/cm^3.

1.2.1 PP-FIBERS, PE-FILMS, AND THEIR WASTES OXIDATION PROCEDURE (IN AN AQUEOUS MEDIUM)

About 1.5 g of the particulate polymer material charged into a round bottom flask equipped with reflux condenser, thermometer, mechanical stirrer and a bubbler for air supplying. 120 mL of H_2O_2 solution and the necessary amount of $FeSO_4 \times 7\ H_2O$ was added, then heated under stirring to 70–90 °C and supplied by air from the compressor during 3–20 h. After the process gone, hydrogen peroxide solution was drained, sample thoroughly washed by water (first tap, then distilled) and dried in air at room temperature.

1.2.2 PP-FIBERS, PE-FILMS, AND THEIR WASTES SOLID-PHASE OXIDATION PROCEDURE

A weighed sample of the particulate polymer material placed into the thermo stated reactor (a glass tube with diameter of 15 mm and a length of 150 mm). The reactor was fed by heated air from the compressor, at a rate of 7.2 L/h. Oxidation was conducted at 85 °C for 4–10 h.

1.2.3 PP-FIBER PE-FILMS AND THEIR WASTES OZONATION PROCEDURE

1. About 1.5 g of the particulate polymer material charged into the thermo stated reactor (a glass tube with diameter of 15 mm and a length of 150 mm). For 60 min ozone-oxygen mixture, produced by an ozonizer (design is described in Ref. [5]), with flow rate of 30 L/h at ambient temperature was passed through the sample. Ozonator's performance by ozone is 12.5 mmol/h. After the ozonation sample is purged with an inert gas to remove the residual ozone and analyzed for HP-groups content.

2. About 1.5 g of the particulate polymer material charged into a round bottom flask equipped with thermometer, mechanical stirrer and a bubbler for air supplying. 120 mL of distilled CCl_4 is added, the ozone-oxygen mixture is passed through the reaction mixture for 60 min at a flow rate of 30 L/h at room temperature. After the ozonation sample is purged with an inert gas to remove the residual ozone, solvent is merged. The sample was washed with chloroform and distilled water, dried in air at room temperature and analyzed for HP-groups content.

1.3 RESULTS AND DISCUSSION

The literature describes the oxidation of the PP-fibers in the medium of toluene by air at a temperature of 70–120 °C in the presence of radical initiators and without them. In practice, these processes lead to the pollution of atmosphere by organic solvents vapor and require the inclusion of the solvent regeneration step in the production scheme. Previously [6], we describe the method of oxidation of the waste PP-fibers by air oxygen in an aqueous medium in the presence of H_2O_2/Fe^{2+} initiating system. This method is also applied to the surface oxidation of the PE-film, and found that when carrying out the oxidation process at 85 °C for 4 h in the presence of 2.7 mol/L H_2O_2 the content of HP-groups in the oxidized material is 0.38×10^{-5} mol/sm^2 (Table 1.1).

Uninitiated solid-phase surface oxidation of PP-fiber, PP- and PE-film with atmospheric oxygen at a temperature of 85 °C was studied for the first time. Found that the content of HP-groups on the surface of oxidized samples, achieved during the process for 4 h, is 3.42×10^{-5} mol/sm^2 for PP-fibers and 1.86×10^{-5} mol/sm^2 for PE-film (Table 1.1).

The most effective method of the oxidation is ozonation that even at room temperature during the process for 1hour, gives the same results as in the previous cases: accumulation of HP-groups on the surface of the studied samples (Table 1.1).

The oxidation conditions and properties of the obtained samples are shown in Table 1.1.

Kinetic curves of HP-groups accumulation during solid-surface oxidation of waste PP-fiber are shows in Figs. 1.2 and 1.3.

TABLE 1.1 Terms of Oxidation of the Waste Polyolefins and Their Properties

Object	Oxidation terms					HP-groups concentration in the material, 10^{-5} mol/sm^2	Limited wetting angle	A, cm^3/g
	Medium	Oxidant	Initiator	Time, hour	T, °C			
PP-fiber (GOST 6574–85)	–	–	–	–	–	0.45	84°	1.10
Waste PP-fiber	–	–	–	–	–	1.5	81°	1.30
Waste PP-fiber	–	air	–	4	85	3.4	69°	1.38
Waste PP-fiber	H_2O	air	2,7 MH_2O_2, 0.37 mg/mLFeSO$_{4x}$7H$_2$O	4	85	3.2	71°	1.39
Waste PP-fiber	–	O_3/O_2	–	1	20	3.3	71°	1.39
PE-film (GOST 17811–78)	–	–	–	–	–	0.31	87°	1.08
Waste PE-film	H_2O	air	2.7 MH_2O_2 0,37 mg/mLFeSO$_{4x}$7H$_2$O	4	85	0.38	69°	1.37
Waste PE-film	–	air	–	4	85	1.8	72°	1.49
Waste PE-film	–	O_3/O_2	–	1	20	2.8	72°	1.49

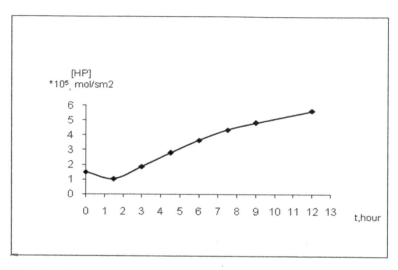

FIGURE 1.2   The dependences of the HP-groups concentration (mol/sm²) on the surface of the waste PP-fiber from the oxidation time (85 °C).

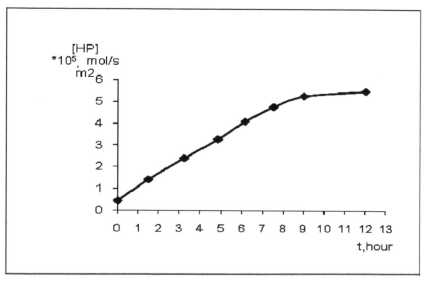

FIGURE 1.3   Dependences of the HP-groups concentration (mol/sm²) on the surfaces of PP-fibers (GOST 26574–26585) from the oxidation time (85 °C).

As follows from the data presented in Fig. 1.1, the storage character of HP groups on the surface of the original PP-fiber having no exploitation, and accordingly, the period of aging in vivo and on the surface of the waste PP-fibers, subjected to aging in vivo, generally, identical. Only the initial part of the kinetic curve of recently produced PP-fibers has the minimum. It can be caused by an intensive breakage of HP-groups emerged in the material during its processing, which dominates over the accumulation of HP-groups during thermal oxidation. It must be concluded that the background of polyolefin material has virtually no effect on the course of its oxidative modification.

Dye ability by azo-dyes and wet ability of the modified samples are studied to confirm the changing of their surface properties. Dye ability of all oxidized samples obtained from waste PP-fiber, PP- and PE-films by azo-dyes is significantly better than dye ability of unmodified materials. Data on reducing of the limited wetting angle of the modified samples indicates increasing of hydrophilic surface. For example, the limited wetting angle of the oxidized waste PE-film surface by water is 10–12° less than the original films.

Also, sorption activity of the modified and unmodified samples measured to evaluate the properties of the polymeric materials obtained by the oxidative modification, (Table 1.1). As can be seen from Table 1.1, the materials subjected to the oxidative modification (aging or oxidation by atmospheric oxygen in an aqueous medium) have greater sorption capacity for water vapor than an unmodified ones. This is probably due, first, to the accumulation of oxygen-containing groups and, respectively, with increasing of hydrophilic properties of the surface, and secondly, with the change in the form of super molecular features on the surface of the polymer macromolecules, leading to the softening of the structure.

1.4 CONCLUSIONS

Thus, comparing the results of oxidative surface modification (oxidation by air oxygen in an aqueous medium or in gas phase, ozonation) of waste fibrous and film materials of PP and PE, we can see that the obtained modified samples has better adhesion and absorption properties than unmodified ones. So we can suggest that there is possibility of their usage as multifunctional additives in various composite materials, polymer-bitumen compositions, etc.

KEYWORDS

- **Chemical modification**
- **Surface oxidation**
- **Waste of polypropylene and polyethylene**

REFERENCES

1. Povstugar, V. I., Kondolopov, V. I. & Mikhailov, S. S. (1988). The Structure and Surfaces Properties of Polymeric Materials, Moscow, Khimiya, 190p.
2. Antonovskii, V. L., & Buzlanova, M. M. (1978). Analytical Chemistry Organics Peroxides Compounds, Moscow Khimiya.
3. Putilova, I. N. (1981). Guide to Practical Work on Colloidal Chemistry, Moscow Visshaya Shkola, 292p.
4. Keltsev, N. V. (1984). Fundamentals of Adsorption Technology, Moscow, Khimiya, 595p.
5. Vendillo, V. G., Emelyanov, Y. M., & Filippov, Y. (1950). The Laboratory Setup for Ozone, Zavodskaya Laboratoriya, *11,* 1401.
6. Bazunova, M. V., Kolesov, S. V., & Korsakov, A. V. (2006). *Journal of Applied Chemistry, 79(5),* 865–867.

CHAPTER 2

A RESEARCH NOTE ON CREATION CARBON-POLYMER NANOCOMPOSITES WITH POLYETHYLENE AS A BINDER

SERGEI V. KOLESOV, MARINA V. BAZUNOVA, ELENA I. KULISH, DENIS R. VALIEV, and GENNADY E. ZAIKOV

CONTENTS

ABSTRACT

A new approach of obtaining of the molded composites on the basis of the mixtures of the powders of nano-dispersed polyethylene, cellulose and the ultra-dispersed carbonic materials is developed. These materials possess the assigned sorption properties and the physic-mechanical characteristics. They are suitable for the usage at the process of cleaning and separation of gas mixture.

2.1 INTRODUCTION

In solving problems of environmental protection, medicine, cleaning and drying processes of hydrocarbon gases are indispensable effective sorbents, including polymer nano composites derived from readily available raw materials.

The nature of the binder and the active components, and molding conditions are especially important at the process of sorption-active composites creating. These factors ultimately exert influence on the development of the porous structure of the sorbent particles and its performance. In this regard, it is promising to use powders of various functional materials having nanoscale particle sizes at the process of such composites creating. First, high degree of homogenization of the components facilitates their treatment process. Secondly, the high dispersibility of the particles allows them to provide a regular distribution in the matrix, whereby it is possible to achieve improved physical and mechanical properties. Third, it is possible to create the composites with necessary sorption, magnetic, dielectric and other special properties combining volumetric content of components [1].

Powders of low density polyethylene (LDPE) prepared by high temperature shearing (HTS) used as one of prospective components of the developing functional composite materials [2, 3].

Development of the preparation process and study of physicochemical and mechanical properties of sorbents based on powder mixtures of LDPE, cellulose (CS) and carbon materials are conducted. As the basic sorbent material new ultrafine nano carbon (NC) obtained by the oxidative condensation of methane at a treatment time of 50 min (NC1) and 40 min (NC2) having a specific surface area of 200 m^2/g and a particle size of 30–50 nm is selected [4]. Ultrafine form of NC may give rise to technological difficulties, for example, during regeneration of NC after using in gaseous environments, as well as during effective separation of the filtrate from the carbon dust particles. This imposes restrictions on the using of NC as an independent sorbent. In this connection, it should be included in a material that has a high porosity. LDPE and CS powders have great interest for the production of such material. It is known that a mixture of LDPE and CS powders have certain absorption properties, particularly, they were tested as sorbents for purification of water surface from petroleum and other hydrocarbons [5].

Thus, the choice of developing sorbents components is explained by the following reasons:

1. LDPE has a low softening point, allowing conducting blanks molding at low temperatures. The very small size of the LDPE particles (60 to 150 nm) ensures regular distribution of the binder in the matrix. It is also important that the presence of binder in the composition is necessary for maintaining of the material's shape, size, and mechanical strength.

2. Usage of cellulose in the composite material is determined by features of its chemical structure and properties. CS has developed capillary-porous structure, that's why it has well-known sorption properties [5] towards polar liquids, gases and vapors.

3. Ultrafine carbon components (nano carbon, activated carbon (AC)) are used as functionalizing addends due to their high specific surface area.

2.2   EXPERIMENTAL PART

Ultrafine powders of LDPE, CS and a mixture of LDPE/CS are obtained by high temperature shearing under simultaneous impact of high pressure and shear deformation in an extrusion type apparatus with a screw diameter of 32 mm [3].

Initial press-powders obtained by two ways. The first method is based on the mechanical mixing of ready LDPE, CS and carbon materials' powders. The second method is based on a preliminary high-shear joint grinding of LDPE pellets and sawdust in a specific ratio and mixing the resulting powder with the powdered activated carbon (БАУ-А mark) and the nano carbon after it.

Composites molding held by thermo baric compression at the pressure of 127 kPa.

Measuring of the tablets strength was carried out on the automatic catalysts strength measurer ПК-1.

The adsorption capacity (A) of the samples under static conditions for condensed water vapor, benzene, n-heptanes determined by method of complete saturation of the sorbent by adsorbate vapor in standard conditions at

20 °C [6] and calculated by the formula: A=m/(Md), where in m – mass of the adsorbed benzene (acetone, n-heptanes), g; M – mass of the dried sample, g; d – density of the adsorbate, g/cm³.

Water absorption coefficient of polymeric carbon sorbents is defined by the formula: $\kappa = \frac{m_{absorbed\,water}}{m_{sample}} \times 100\%$, wherein $m_{absorbed\,water}$ is mass of the water, retained by the sorbent sample, m_{sample} is mass of the sample.

Experimental error does not exceed 5% in all weight methods at P = 0.95 and the number of repeated experiments n = 3.

2.3   RESULTS AND DISCUSSION

Powder components are used as raw materials for functional composite molding (including the binder LDPE), because molding of melt polymer mixtures with the active components has significant disadvantages. For example, the melt at high degrees of filling loses its fluidity, at low degrees of filling flow rate is maintained, but it is impossible to achieve the required material functionalization.

It is known that amorphous-crystalline polymers, which are typical heterogeneous systems, well exposed to high-temperature shear grinding process. For example, the process of HTS of LDPE almost always achieves significant results [3]. Disperse composition is the most important feature of powders, obtained as result of high-temperature shear milling. Previously, on the basis of the conventional microscopic measurement, it was believed that sizes of LDPE powder particles obtained by HTS are within 6–30 micrometers. Electron microscopy gives the sizes of 60 to 150 nm. The active powder has a fairly high specific surface area (upto 2.2 m²/g).

The results of measurement of the water absorption coefficient and of the static capacitance of LDPE powder by n-heptanes vapor are equal to 12% and 0.26 cm³/g, respectively. Therefore, the surface properties of LDPE powder more developed than the other polyethylene materials'.

Selection of Molding Conditions of Sorbents Based on Mixtures of LDPE, CS and Ultrafine Carbon Materials' Powders

Initial press-powders obtained by two ways. The first method is based on the mechanical mixing of ready LDPE, CS and carbon materials' powders. The second method is based on a preliminary high-shear joint grinding of

LDPE pellets and saw dust in a specific ratio and mixing the resulting powder with the powdered activated carbon and the nano carbon after it. The methods of moldings thermo baric pressing at pressure of 127 kPa.

The mixture of LDPE/CS compacted into cylindrical pellets at a temperature of 115–145 °C was used as a model mixture for selection of composites molding conditions. Pressing temperature should be such that the LDPE softens but not melts, and at the same time forms a matrix to prevent loss of specific surface area in the ready molded sorbent due to fusion of pores with the binder. The composites molded at a higher temperature, have a lower coefficient of water absorption than the tablets produced at a lower temperature, that's why the lowest pressing temperature (120 °C) is selected. At a higher content of LDPE the water absorption coefficient markedly decreases with temperature.

Cellulose has a high degree of swelling in water (450%) [5], this may lead to the destruction of the pellets. Its contents in samples of composites, as it has been observed by the sorption of water, should not exceed 30 wt%. There is a slight change of geometric dimensions of the pellets in aqueous medium at an optimal value of the water absorption coefficient when the LDPE content is 20 wt%.

Samples of LDPE/CS with AC, which sorption properties are well studied, are tested for selecting of optimal content of ultrafine carbon. The samples containing more than 50 wt% of AC have less water absorption coefficient values. Therefore, the total content of ultrafine carbon materials in all samples must be equal to 50 wt%.

Static capacitance measurement of samples, obtained from mechanical mixtures of powders of PE, CS and AC, conducted on vapors of n-heptanes and benzene, to determine the effect of the polymer matrix on the sorption properties of functionalizing additives. With a decrease of the content of AC in the samples with a fixed (20 wt%) amount of the binder, reduction of vapor sorption occurs. It indicates that the AC does not lose its adsorption activity in the composition of investigated sorbents.

Strength of samples of sorbents (Fig. 2.1) is in the range of 620–750N. The value of strength is achieved in the following molding conditions: t = 120 °C and a pressure of 127 kPa.

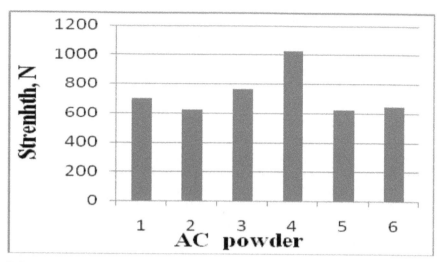

FIGURE 2.1   Comparison of strength of pellets, based on LDPE, CS (different species of wood) and AC powders.1 – sorbent of LDPE/AC/CS = 20/50/30wt% based on the powders of jointly dispersed pellets of LDPE and softwood sawdust with subsequently addition of AC; 2 – sorbent of LDPE/AC/CS = 20/50/30 wt% based on the powders of jointly dispersed pellets of LDPE and hardwood sawdust with subsequently addition of AC; 3 – sorbent of LDPE/AC/CS = 20/50/30 wt% based on the mechanical mixtures of the individual powders of LDPE, CS from softwood and AC; 4 – AC tablet; 5– sorbent of LDPE/CS = 20/80 wt%; 6 – sorbent of LDPE/AC = 20/80 wt%.

Thus optimal weight compositions of the matrix of LDPE/CS compositions 20/30wt% with 50wt% containing of carbon materials.

Sorption Properties of Carbon-Polymer Composites by Condensed Vapors of Volatile Liquids

For a number of samples of sorbents static capacitance values by benzene vapor is identified (Fig. 2.2). They indicate that the molded mechanical mixture of 20/25/25/30 wt% LDPE/AC/NC1/CS has a maximum adsorption capacity that greatly exceeds the capacity of activated carbon. High sorption's capacity values by benzene vapor appear to be determined by weak specific interaction of π-electron system of the aromatic ring with carboxylic carbon skeleton of the nano carbon [7].

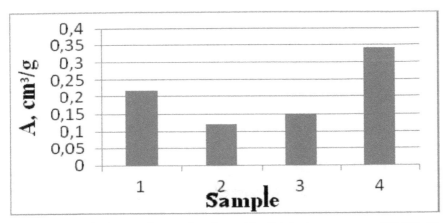

FIGURE 2.2   Static capacitance of sorbents A(cm³/g) by benzene vapor (20 °C). (1) Molded mechanical mixture of LDPE/AC/NC1/CS= 20/25/25/30wt%; (2) Molded mechanical mixture of LDPE/AC/NC2/CS = 20/25/25/30 wt%; (3) Molded mechanical mixture of LDPE/AC/CS=20/50/30wt%; (4) AC medical tablet (controlling).

Static capacitance of obtained sorbents by heptanes vapors significantly inferiors to capacity of activated carbon (Fig. 2.3); probably it is determined by the low polarizability of the molecules of low-molecular alkenes. Consequently, the investigated composites selectively absorb benzene and can be used for separation and purification of mixtures of hydrocarbons.

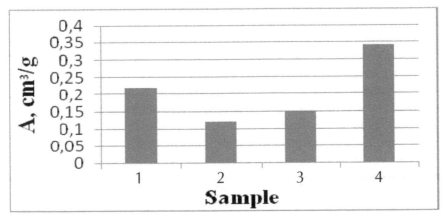

FIGURE 2.3   Static capacitance of sorbents A(cm³/g) by n-heptanes vapor (20 °C). (1) Molded mechanical mixture of LDPE/AC/NC1/CS= 20/25/25/30wt%; (2) Molded mechanical mixture of LDPE/AC/NC2/CS = 20/25/25/30 wt%; (3) Molded mechanical mixture of PE/AC/CS=20/50/30 wt%; (4) AC medical tablet (controlling).

Molded composite based on a mechanical mixture of LDPE/AC/NC1/CS = 20/25/25/30 wt% has a sorption capacity by acetone vapor comparable with the capacity of activated carbon (0.36 cm³/g) (Fig. 2.4).

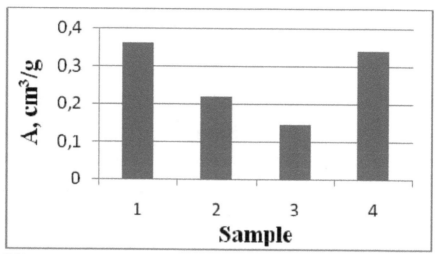

FIGURE 2.4   Static capacitance of sorbents, A (cm³/g) acetone vapor (20 °C). (1) molded mechanical mixture of LDPE/AC/NC1/CS= 20/25/25/30wt.%; (2) molded mechanical mixture of LDPE/AC/NC2/CS = 20/25/25/30 wt. %; (3) molded mechanical mixture of LDPE/AC/CS=20/50/30 wt%; (4) AC medical tablet (controlling).

Sorbents' samples containing NC2 have low values of static capacity by benzene, heptanes and acetone vapor. It can be probably associated with partial occlusion of carbon material pores by remnants of resinous substances by products of oxidative condensation of methane, and insufficiently formed porous structure.

The residual benzene content measuring data (Table 2.1) shows that the minimal residual benzene content after it's desorption from the pores at t = 70 °C for 120 minutes observes in case of sorbent LDPE/AC/NC1/CS composition = 20/25/25/30 wt%. It allows concluding that developed sorbents have better ability to regenerate under these conditions in comparison with activated carbon.

TABLE 2.1 Sorbents' Characteristics Total Pore Volume V_{tot}; Static Capacitance (A) by Benzene Vapors at the Sorption Time of 2 days; Residual Weight of the Absorbed Benzene after Drying at t=70 °C for 120 min.

LDPE/AC/NC/CS sorbent composition, wt. %	$V_{tot.}$, cm³/g	A, cm³/g	Residual benzene content as a result of desorption, %
20/25/25/30	1.54	0.5914	2.9
20/50/–/30	1.21	0.1921	10.3
–/100/–/–	1.60	0.3523	32.0

2.4 CONCLUSIONS

Thus, the usage of nano sized LDPE as a binder gives a possibility to get the molded composite materials with acceptable absorption properties. Optimal conditions for molding of sorbents on the basis of mixtures of powdered LDPE, cellulose and ultrafine carbon materials were determined: temperature 120 °C and pressure of 127 kPa, content of the binder (polyethylene) is 20 wt%.

Varying the ratio of the components of the compositions on the basis of ternary and quaternary mixtures of powdered LDPE, cellulose and ultrafine carbon materials it is possible to achieve the selectivity of sorption properties by vapors of certain volatile liquids. Established that molded mechanical mixture of LDPE/AC/NC1/CS 20/25/25/30 wt% has a static capacity by condensed vapors of benzene and acetone 0.6 cm³/g and 0.36 cm³/g respectively, what exceeds the capacity of activated carbon. The static capacitance of the compositions by the n-heptanes vapors is 0.21 cm³/g; therefore, the proposed composites are useful for separation and purification of gaseous and steam mixtures of different nature.

Developed production method of molded sorption-active composites based on ternary and quaternary mixtures of powdered LDPE, cellulose and ultrafine carbon materials can be easily designed by equipment and can be used for industrial production without significant changes.

KEYWORDS

- Cellulose
- High-temperature shift crushing
- Nano-carbon
- Polyethylene
- Sorbents

REFERENCES

1. Akbasheva, E. F., & Bazunova, M. V. (2010). Tableted Sorbents Based on Cellulose Powder Mixtures, Polyethylene and Ultra-Dispersed Carbon, *Materials Open School Conference of the CIS "Ultrafine and Nano Materials" (11–15 October 2010)*, Ufa: Bashkir State University, 106.
2. Enikolopyan, N. S., Fridman, M. L., & Karmilov, A. Yu. (1987). Elastic-Deformations Grindings of Thermo-Plastics Polymers, Reports AS USSR, *296(1)*, 134–138.
3. Akhmetkhanov, R. M., Minsker, K. S., & Zaikov, G. E. (2006). On the Mechanisms of Fine Dispersions of Polymers Products at Elastic Deformations Effects, *Plasticheskie Massi, 8*, 6–9.
4. Aleskovskiy, V. B., & Galitseisky, K. B. (20.11.2006). Russian Patent "Method of Ultrafine Carbon" *2287543*.
5. Raspopov, L. N., Russiyan, L. N., & Zlobinsky, Y. I. (2007). Waterproofs Composites Comprising Wood and Polyethylene Dispersion, *Russian Polymer Science Journal, 50(3)*, 547–552.
6. Keltsev, N. V. (1984). *Fundamentals of Adsorptions Technology*, Moscow Chemistry, 595p.
7. Valinurova, E. R., Kadyrov, A. D., & Kudasheva, F. H. (2008). Adsorption Properties of Carbon Rayon, *Vestn Bashkirs Univer, 13(4)*, 907–910.

CHAPTER 3

A RESEARCH NOTE ON THE INFLUENCE OF HYBRID ANTIOXIDANTS ICHPHANS ON THE STRUCTURE OF LIPOSOME LIPID BILAYER

E. YU. PARSHINA, L. YA. GENDEL, and A. B. RUBIN

CONTENTS

ABSTRACT

Analysis of the kinetics of a scorhate-induced reduction of radical centers in liposome-incorporated spin probes revealed that Ichphans could modify different regions of the liposome structure depending on their hydrophobic properties and intra membrane distribution. Thus, the membranotropic action of Ichphans may involve a lipid component.

3.1   INTRODUCTION

Hybrid antioxidants Ichphans [l] comprise fragments with antioxidant and ant cholinesterase activities [2–4] and an aliphatic substituent at the quaternary nitrogen determining their hydrophobicity (Table 3.1). The combined biological activity of these compounds makes them promising drugs for treating Alzheimer's disease [2–4]

We have shown that intercalation of Ichphans into the erythrocyte membrane and their distribution in the intra membrane space are accompanied by changes in the cell morphology and plasma membrane structure [5–7].Membrane transport of Ichphans and their influence on membrane (particularly, lipid bi layer) structure may contribute to their biological effects by altering, among other things, the activity of membrane-bound and lipid-dependent enzymes.

The influence of Ichphans on lipid bi layer structure has not been studied heretofore. The aim of this work was to assess this aspect with liposome, spin probes, and a series of Ichphans of varied hydrophobicity.

3.2   EXPERIMENTAL PART

The homologous Ichphan series [I] is shown in the (Table 3.1). Liposomes were prepared by sonicating L-α-lecithin (Sigma) in 50 mM K-phosphate pH 7.5 with 2 mM $MgSO_4$ in a UZDN-2T device.

The spin probes were (I) 5-doxylstearate (Sigma) and (II) a γ-carboline derivative [8] (structures shown in Table 3.1).

Ichphans and probes were dissolved in ethanol (within 1.2%vol in the sample).

EPR spectra were taken with an RE-1306 instrument at 18±2 °C. Statistical processing was performed using the Student's t-test.

3.3 RESULTS AND DISCUSSION

The effect of Ichphans on the liposome membrane was assessed from the kinetics of ascorbate-induced reduction of the radical centers of membrane-incorporated spin probes (EPR signal decay) [9].

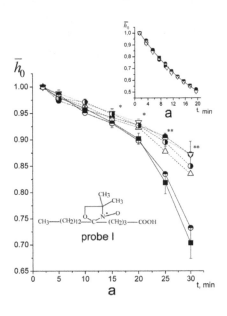

a

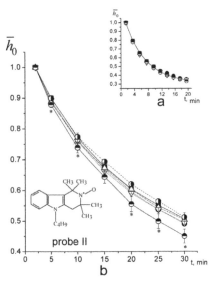

b

Influence of Ichphan homolog on adsorbate-induced reduction of the radical centers of spin probes I (left) and II (right) in ethanol (a) and in the liposomal membrane (b). Symbols: (filled square) control, (empty square) I(C-I), (triangle) I(C-8), (turned triangle) I(C-l0), (filled circle) I(C-12), (empty circle) I(C-16); $\bar{h}_0 = h_0(t)/h_0(t=2\min)$, where $h_0(t)$ is the intensity of the central component in the nitroxyl EPR spectrum as a function of time t after addition of ascorbate. Concentrations: 40 mg/mL lecithin, 4 mM ascorbate, 0.1 mM probes, 0.5 mM Ichphan. Asterisks mark significant difference from control $(p < 0.05)$.

TABLE 3.1 Ichphans

Structural formula	X	R	Designation
	J⁻	$-CH_3$	I(C-1)
H3C, CH3 OH CH3 ... CH2CH2COOCH2CH2—N⁺—R X⁻ Br		$-C_8H_{17}$	I(C-8)
		$-C_{10}H_{21}$	I(C-10)
		$-C_{12}H_{25}$	I(C-12)
		$-C_{16}H_{33}$	I(C-16)

A major factor influencing the radical reduction (i.e., EPR signal decay) rate is the structural state of the membrane, which determines the rate of adsorbate anion diffusion into the region of intra membrane probe distribution.

Control assays in panel (a) of the figure show that spin probes I and II are reduced by adsorbate in ethanol at different rates, whereas Ichphans themselves have no effect at all in this system panel (b) demonstrate that in the liposome membrane, as in ethanol, probe II is reduced at a higher rate. Reduction of probe I in the membrane is inhibited to a varying extent by Ichphans with 1 to 12 carbons in the hydrocarbon substituent while I(C-16) with the longest R is in effective (panel Ib). Conversely, in the case of probe II this compound appreciably accelerated the reduction, while I(C-1)-I(C-12) had no reliable effect (panel IIb).

Hence, Ichphans can modify the structure of the lipid bi layer so as to hinder or promote the access of the reducer to the probe. The opposite effects observed for the two probes perhaps indicate that these probes localize in different membrane regions. Likewise, clear grouping of Ichphan homolog

(1–12 versus 16) in their effect perhaps means that the hydrocarbon chain length is essential for their partitioning within the membrane. One may think that I(C-1)-I(C-12) penetrate the membrane interior according to their hydrophobicity and membrane capacity, whereas the bulkiest I(C-16) remains near the membrane surface, closely to probe II.

Thus, the membranotropic action of Ichphans may involve a lipid component.

KEYWORDS

- **Anti cholinesterase drugs**
- **Hybrid compounds**
- **Liposomes**
- **Membrane transport**
- **Spin probes**
- **Synthetic antioxidants**

REFERENCES

1. Nikiforov, G. A., Belostotskaya, I. S., Vol'eva, V. B. et al. (2003). Nauchnyi Vestnik Tyumenskoi Meditsinskoi Akademii, Spetsial'nyi Vypusk "Biooksidanty," (Sci. Herald of Tyumen Medical Academy: Special Issue "Biooxidants" 50–51.

2. Braginskaya, F. I., Molochkina, E. M., Zorina, O. M. et al. (1996) In *Alzheimer Disease From Molecular Biology to Therapy,* Ed. by R. Becker and E. Giacobini (Birkhauser, Boston), 337–342.

3. Molotchkina, E. M., Ozerova, I. B., & Burlakova, E. B. (2002). Free Radical Biology and Medicine, 33, Issue 2S1, *610*, 229–230.

4. Burlakova, E. B., Molochkina, E. M., Goloshchanov, A. N., et al. Dec. 10–11–2003. In Abstracts of the Conference "Basic Research for Medicine," Moscow, 15–17.

5. Parshina, E. Yu., Gendel, L.Ya, & Rubin, A. B. (2011). In "Progress in Study of Chemical and Biochemical Reactions" *Kinetics and Mechanism*, 71–78.

6. Parshina, E. Yu., Gendel, L. Ya., Rubin, A. B. (2012). *Pharmaceutical Chemistry Journal, 46(2),* 82–86.

7. Parshina, E. Yu., Gendel, L. Ya., & Rubin, A. B. (2013). *Polymers Research Journal, 7(7),* 361–367.

8. Panasenko, O. M., Gendel, L. Ya., & Suskina, V. I. (1980). Izv Akad Nank SSSR, *Ser Biol.,* 6, 854–864.

9. Gendel, L. Ya., & Kruglyakova, K. E. (1986). In *Spin Labeling and Probing Method* (Nauka, Moscow), 163–194 (in Russian).

CHAPTER 4

DYNAMICALLY VULCANIZED THERMOELASTOPLASTICS BASED ON BUTADIENE-ACRYLONITRILE RUBBER AND POLYPROPYLENE MODIFIED NANOFILLER

S. I. VOLFSON, G. E. ZAIKOV, N. A. OKHOTINA,
A. I. NIGMATULLINA, O. A. PANFILOVA, and A. A. NIKIFOROV

CONTENTS

ABSTRACT

Thermoplastic vulcanizates based on polypropylene and nitrile-butadiene rubber, containing modified organoclay were developed. It was shown that composites containing 1 to 5 pbw of Cloisite 15A montmorillonite added to rubber show improved physical-mechanical characteristics. Their swelling degree in AI-92 and motor oil was determined. The swelling degree of composites in petrol and motor oil decreases substantially, by 20–63%, due to the introduction of Cloisite15A montmorillonite. Modification of thermoplastic vulcanizates using layered silicates raised the degradation onset temperature and decreases weight loss upon high temperature heating.

4.1   INTRODUCTION

Polymer nanocomposite are the most effective advanced materials for different areas of application any researches of polymer based on reinforced nanocomposite, which are very perspective direction of modern polymer science. It is well known [1, 2], that the polymers filled with small quantities (about 5–7 pbw) nanoparticles demonstrate great improvement of thermo mechanical and barrier properties. The most part of published works is devoted to polyolefin's filled with nanoparticles of laminated silicates, mainly montmorillonite (MMT).

The production and application of dynamically vulcanized thermoelastoplastics (DTEP, TPV) are intensively developed in the last years. That is connected with the fact, that TPV combined the properties of the vulcanized elastomer during utilization and thermoplastics in the processing. By their characteristics, the method of manufacturing, as well as processing TPV are in principle significantly differing from both plastics and elastomer. Due to their high level of properties the production of DTEP increases approximately by 10–15% per year in the world. Dynamic thermoelastoplastics elastomers are complicated composition materials consisting of continuous phase of thermoplastic and discrete phase of cured elastomer [3–9].

Dynamically vulcanized thermoelastoplastics based on butadiene-acrylonitrile elastomer (NBR) and polypropylene (PP) possesses the special interest for production of car parts and petroleum industry. At the same time low compatibility of components during mixing of polar elastomer phase and nonpolar thermoplastic phase in pair NBR–PP is observed. That leads to decreased elastic-strength and thermal properties of composition material of this type. The effective ways of improvement TPV properties is the use of natural nano sizes filler oreganos clay particularly montmorillonite. Montmorillonite could create organo mineral complexes due to liability of laminated structure of MMT swelling during intercalation of organic substances. So, it was very interesting to investigate the influence of small quantity of modified montmorillonite clay on DTEP properties.

4.2 EXPERIMENTAL PART

Butadiene-acrylonitrile elastomer with 18% content acrylonitrile (NBR-18) and polypropylene Balen grade 01030 were used for manufacturing of DTEP in mixing camera of the plastic order Brabender at 180 °C during 10 min [10–12]. Ratio rubber-polypropylene was 70:30 and 50:50 in the TPV, the sulfur-based vulcanizing system was used. As a filler was used MMT Cloisite15A brand of USA Rockwood company with variable capacity 125 ekv/100 g, modified by quaternary ammonium salts $[(RH)_2(CH_3)_2N]^+Cl^-$ where R–is a residue of hydrogenated fatty acids C_{16}–C_{18}. The content of MMT varied from 1 to 3 pbw at 100 pbw of polymer phase MMT was preliminarily added into elastomer or into polypropylene.

The structural characteristics of MMT were determined by powder X-ray method at diffractometer D8ADVANCE Brucker Company in Breg-Brentano in the mode of the step scanning. Mono chromating radiation was performed by anticathode with wavelength $\lambda = 0.154178$ nm. The X-ray tube action mode was 40 kV and 30 mA. The size of the slit was $1 \times V$ 20.

The elastic-strength characteristics by device Inspect mini of Monsanto Company at 50 mm/min speed were determined.

The elastic-hysteresis properties of modified DTEP were estimated by dynamic rheometer RPA 2000 Alfa technologic company (USA).

During the test cycle shift deformation of the sample was performed. The torsional factor G* by use of Fourier transformation was divided in two components: true G' and virtual G," which characterize elastic and plastic properties of the material consequently. The м modulus of elasticity S' and loss modulus S," the same as the ratio of plastic and elastic components (tg δ) were also determined by RPA 2000 device.

The tests of modified dynamic thermoplastic elastomer by RPA 2000 were performed in two modes: at constant temperature 100 °C, constant frequently of deformation 1 Hz and variable of deformation from 1 to 10%; at constant temperature 100 °C, constant of deformation 1% and variable deformation frequently from 0.1 to 10 Hz.

The thermal behavior of composites was investigated by simultaneous thermo analyzing device STA 409 PC NETZSCH Company by thermogravimetric method (TG).

The swelling degree was estimated in liquid aggressive media (such as AI-92 petrol or motor oil) during 72 h at various temperatures (depending on medium type): 23, 70, and 125 °C;

4.3   RESULTS AND DISCUSSION

The investigation of the DTEP structural characteristics demonstrated the whole series reflexes with d~3.3 (the strongest reflex) 1.6–1.1 nm, which is typical for MMT Cloisite15A on the diffract gram of DTEP sample with 3 pbw MMT. At the same time the reflex in small angle area (1.5–3° 2q) with inter planar distance 4.2–4.8 nm for DTEP with 1 pbw MMT was observed.

This reflex is the first basal diffraction of organomontmorillonite from the second basal diffraction. It is characterized by slight increase of the background and third reflex is absent in this case. It is connected with exfoliation MMT in polymer matrix. And we can forecast the significant improvement of physical-mechanical and thermal properties of DTEP in this case.

Physical-mechanical characteristics of composites are presented in Table 4.1.

TABLE 4.1   Elastic-Strength Properties of MMT-Modified DTEP

Properties	Content MMT, pbw		
	0	1	3
Mixing MMT with PP			
DTEP 70:30			
Conditional tensile strength, MPa	3.72	4.24	4.14
Relative elongation, %	140	170	160
Modulus of elasticity, MPa	93	118	100
DTEP 50:50			
Conditional tensile strength, MPa	8.2	8.8	8.4
Relative elongation, %	119	168	153
Modulus of elasticity, MPa	170	285	200
Mixing MMT with rubber			
DTEP 70:30			
Conditional tensile strength, MPa	3.72	4.32	4.27
Relative elongation, %	140	180	170
Modulus of elasticity, MPa	93	143	131
DTEP 50:50			
Conditional tensile strength, MPa	8.2	9.0	8.5
Relative elongation, %	119	172	163
Modulus of elasticity, MPa	170	329	292

As seen from Table 4.1 composites containing 1 to 5 pbw of Cloisite 15A montmorillonite added to rubber show improved physical-mechanical characteristics. In comparison with unfilled TPV, the modulus of elasticity increases by factor of 1.94÷2.11, tensile strength a increases by factor of 1.09÷1.11, and relative elongation increases by factor of 1.37÷1.47.

It is known that increase of M modulus of elasticity S' and decrease of loss modulus S" as well as tangent of mechanical losses $tg\ \delta$ should lead to improvement of physical-mechanical properties of polymer materials. The data, received by RPA 2000 method, at the first (Fig. 4.1) and second (Fig. 4.2) modes of testing of modified by MMT DTEP on Figs. 4.1 and 4.2 are presented.

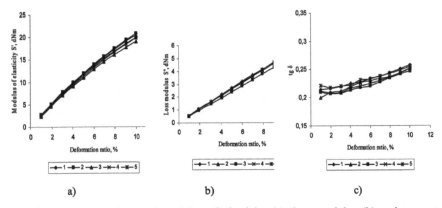

FIGURE 4.1 Dependence of modulus of elasticity (a), loss modulus (b) and tangent of mechanical losses (c) on deformation ratio for DTEP with ratio rubber-polypropylene 70:30 with different content of MMT: 1 – 0 pbw MMT; 2 – 1 pbw MMT in the rubber; 3 – 1 pbw MMT in the polypropylene; 4 – 3 pbw MMT in the rubber; 5 – 3 pbw MMT in the polypropylene.

According to results of these tests, we can propose the increase of elastic-hysteresis properties in the raw DVTEP + 1 pbw MMT in the rubber > DTEP + 1 pbw MMT in the polypropylene > DVTEP + 3 pbw MMT in the rubber > DTEP + 3 pbw MMT in the polypropylene.

The mixing of MMT with PP the same as with butadiene-acrylonitrile rubber leads to improvement of elastic-strength properties of DTEP especially M modulus of elasticity, which increased by 27–54%.

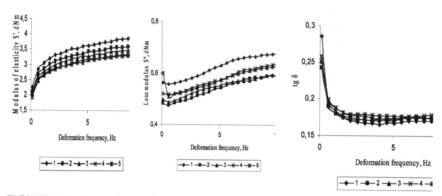

FIGURE 4.2 Dependence of modulus of elasticity (a), loss modulus (b) and tangent of mechanical losses (c) on deformation frequency for DTEP with ratio rubber: polypropylene 70:30 with different content of MMT: 1 – 0 pbw MMT; 2 – 1 pbw MMT in the rubber; 3 – 1 pbw MMT in the polypropylene; 4 – 3 pbw MMT in the rubber; 5 – 3 pbw MMT in the polypropylene.

The best results are achieved at preliminary mixing of MMT with NBR-18. It is important to underline that the less quantity of MMT is used the greater of the possibility of its exfoliation as separate particles in polymer matrix consequently, the improvement of elastic-strength properties was achieved. With increase of filler content agglomerates of greater size are formed. They play the role of structure defects and induce the decrease of conditional strength. The best elastic-strength properties of DTEP, modified by 1 pbw MMT are explained by better distribution of filler in the rubber phase.

Since the materials based on the formulated DTEP are intended to be exposed to higher temperatures, the effect of modified layer silicates as additives on TPV thermal stability was studied. Thermal test results are shown in Table 4.2.

TABLE 4.2 Thermal Behavior of DTEP (70:30 NBR to PP)

Properties	Content MMT, pbw		
	0	**1**	**3**
Mixing MMT with PP			
Degradation onset temperature, °C	269.2	327	314
Mixing MMT with rubber			
Degradation onset temperature, °C	269.2	350.4	343.4

As shown in Table 4.2, modification of DTEP using layered silicates increases the degradation onset temperature and decreases the weight loss upon high temperature heating. The unfilled TPV has the degradation onset temperature of 269 °C. TPV compositions containing 1 and 3 pbw of Cloisite 15A MMT added to rubber show the best thermal stability (350 and 343 °C, respectively). Introduction of layered silicate into DTEP at equal proportions of NBR and PP also results in the degradation onset temperature rising by 30 °C.

Due to DTEP application in automobile and petroleum industries it is important to exam their resistance to aggressive media, such are petrol Euro 3 and motor oil. The examined composites exposed in aggressive media for 72 h at different temperatures and this data are represented in Table 4.3.

TABLE 4.3   Swelling Degree%, of DTEP (70:30 NBR to PP), Modified by Montmorillonite

Conditions	Content MMT, pbw		
	0	**1**	**3**
Mixing MMT with PP			
23°C, petrol	15	5.8	5.6
23°C, motor oil	4.5	3.8	3.4
70°C, motor oil	11.2	4.3	7.1
125°C, motor oil	15.5	6.1	8.6
Mixing MMT with rubber			
23°C, petrol	15	8	7.6
23°C, motor oil	4.5	3.1	3.1
70°C, motor oil	11.2	4.4	5
125°C, motor oil	15.5	6.2	6.5

As can be seen from Table 4.3, the filler significantly decrease the swelling degree and, consequently, increase the resistance to aggressive media.

When MMT is mixed with PP as well as with NBR the best resistance to aggressive media is achieved by use of 1 pbw of additive. The materials are more stable in motor oil, than in petrol. The swelling degree of DTEP is increase with increasing of temperature exposition in motor oil independently on composition of materials.

4.4 CONCLUSIONS

Thermoplastic vulcanizates have been formulated which are based on polypropylene and nitrile-butadiene rubber with a modified organoclay as a filler. It was established that the best physical-mechanical and thermal properties by preliminary mixing of small quantities nano filler with elastomer are achieved. The good correlation of results of dynamic analysis of polymers by RPA 2000 and structural, physical-mechanical, thermal and barrier characteristics of DTEP, modified by MMT was achieved.

KEYWORDS

- **Montmorillonite**
- **Nano filler**
- **Nitrile-butadiene rubber**
- **Organoclay**
- **Physical-mechanical characteristics**
- **Polypropylene**
- **Thermoplastic**
- **Vulcanization**

REFERENCES

1. Volfson, S. I. (2004). Dynamically Vulcanized Thermoplastics Production Processing Properties, Nauka, Moscow.
2. Pinnavaia, T. J., Beall, G. W., John Willey, Sons et al. (2001). Polymer-Clay Nano Composites New York, 349p.
3. Polymer Nano Composites Synthesis, Characterization, and Modeling, Krishnamoorti, R., Vaia, A. R. et al. (2001). American Chemical Society Washington, 242p.
4. Utrachki, L. E. (2004). Clay-Containing Polymeric Nano Composites Monograph to be published by Rapra in 600p.
5. Alexandre, M. (2000). Dubois Ph. Polymer-Layered Silicate Nano Composites Preparation, Properties, and Uses of a New Class of Materials Mater Sci. Eng., *28,* 1–63.
6. Mikitaev, A. C., Kaladzhyan, A. A., Lednev, O. B., & Mikitaev, M. A. (2004). Nanocomposite Polymer Materials Based on Organoclay Plastic Mass, *12,* 45–50.
7. Sheptalin, R. A., Koverzyanova, E. V., Lomakin, S. M., & Osipchik, V. S. (2004). The Features of Inflammability and Thermal Destruction of Nano Composites of Elastic Polyurethane Foam Based on Organic Modified Layered Alumino Silicate, *Plastic Mass, 4,* 20–26.
8. Lomakin, S. M., & Zaikov, G. E. (2005). Polymer Nano Composites with Decreased Inflammability Based on Layered Silicates, *High-Molecular Substance, Series B, 47(1),* 104–120.

9. Mikitaev, A. C., Kaladzhyan, A. A., Lednev, O. B. et al. (2005). Nanocomposite Polymer Materials Based on Organoclay with Increased Fire-Resistance, Plastic Mass, *4*, 36–43
10. Vol'fson, S. I., Nigmatullina, A. I., Sabirov, R. K., Lyzina, T. Z., Naumkina, N. I., & Gubaidullina, A. M. (2010). Effect of Organoclay on Properties of Dynamic Thermoelastoplastics Russian, *Journal of Applied Chemistry, 83(1),* 123–126
11. Nigmatullina, A. I., Volfson, S. I., Okhotina, N. A., & Shaldibina, M. S. (2010). Properties of Dynamically Vulcanized Thermoplastics Containing Modified Polypropylene and Layered Filler, Vestnik of Kazan Technological University, *9,* 329–333.
12. Elastic and Hysteretic Properties of Dynamically Cured Thermoelastoplastics Modified by Nano Filler, Volfson, S. I., Okhotina, N. A., Nigmatullina, A. I., Sabirov, R. K., Kuznetsova, O. A., & Akhmerova, L. Z. (2012). *Plastic Mass, 4,* 42–45.

CHAPTER 5

MODEL REPRESENTATIONS OF THE EFFECT OF TEMPERATURE ON RESISTANCE POLYPROPYLENE FILLED WITH CARBON BLACK

N. N. KOMOVA and G. E. ZAIKOV

CONTENTS

ABSTRACT

The analysis of the relative resistivity changing, crystallization and kinetics of crystallization was carried out using the difference scanning calorimetry and ohmmeter for the carbon black filled samples of polypropylene. Two stages of crystallization were observed. They were described by means of the generalized equation of Erofeev. Obtained by modeling representations the kinetic parameters are in satisfactory agreement with the experimental data.

5.1   INTRODUCTION

Addition of various kinds of fillers is used to make composite materials based on polymers required for engineering properties. One of the fillers applied in high-molecular materials is carbon used in a form of natural carbon black or graphite. The carbon black is an active strengthening material. Its strengthening action is the stronger the smaller are the dimensions and specific surface area and the higher is the surface energy of the particles. Black particles are composed of a large number of agglomerated crystalline elements called micro crystallites. In general, the systems of micro crystallites have the shape of spheres or sintered spheres, which may form spatial branched chains [1]. Forming such chains, aggregates exhibit the properties of fibrous fillers. As a result, the resistance of the composite material filled with carbon black particles is reduced even at a small content of carbon black [2].

Polypropylene (PP) filled with carbon black with content ranging from 10 wt%, is a conductive material. On the other hand, the mechanical properties of the polypropylene filled with carbon black such as strength and impact resistance are satisfactory when the contents of carbon black is less than 20% [2, 3]. Properties of electrical conductivity in the composites are based on polypropylene and the carbon black is determined by the formation of chains of particles conglomerates in the amorphous phase [3–5]. Along with this, in Ref. [6] on the basis of molecular dynamic simulation it was shown that the adsorption energy and the portion of the polymer on the surface of amorphous carbon black are determined by the area of the hydrophobic contact (CH_2 and CH_3 groups) and the presence of electronegative atoms in the structure of the polymer. The conformation of the polymer fragment, or complementarity of its geometry on the surface of carbon black also affects the interaction between the polymer molecules and a solid particle surface. Calculation by quantum mechanical simulation showed that the enthalpy of adsorption to particles of carbon black PP is amount of 24.57 kJ/mol, which is 1.5 higher than for PE [6]. So, we can conclude that the formation of the morphological structure of the polymer matrix is affected by the filler both displaced into amorphous phase and contacted with the filler particles. The temperature effect on each of the factors determining the morphology of the polymer matrix is not unique. Since the morphology of the system significantly affects the electrical conductivity the effect of temperature on the electrical conductivity also has a complex character. In this paper we analyze the relative resistivity change and the degree of crystallinity of the system (PP filled with carbon black) affected by temperature. A description of processes kinetics and estimation of experimental results have been made.

5.2   EXPERIMENTAL PART

5.2.1   MATERIALS

The object of study is a composition based on the brand PP 01050 (TU 2211–015–00203521–99) (density of 900 kg/m³. Melt index 4.0–7.0 g/10 min), with the same content of conductive carbon black (CB) PA-76 (TU 38–10001–94) about 11.7% (20% weight fraction).

Preparation of the compositions was performed in a closed rotary mixer "Brabender" volume of the working chamber with 30 mL of liquid heated for 10 minutes at a rotor speed of 50 rev/min. Mixing temperature was 190°C. The process of sample making for tests was carried out in a hydraulic press at a temperature of 200 °C. Samples were cooled between the steel plates in the mold for 20 minutes from 200 to 70 °C after they were finally cooled in air. Measurement of volume resistivity samples (ρ) at elevated temperatures (T) was carried out in a heat chamber as much as 3 °C/min. Electrical resistance of the samples was measured at a voltage ohmmeter DT9208A 9 V. Melting process electroconductive samples were examined by differential scanning calorimetry (DSC) by TA instrument Pyris 6 DSC Perkin Elmer with heating from 25 to 250 °C at 3 °C/min (as in the study of changes in electrical resistance when heated). The degree of crystallinity was calculated using the equation: $\alpha_{cryst} = \Delta H_{melt}/\Delta H_{melt0}$, where ΔH_{melt} – enthalpy of polymer fusion phase of the sample calculated from the results of DSC based on the mass fraction of carbon, ΔH_{melt0} – melting enthalpy of the polymer with a 100% crystallinity [8].

5.3   RESULTS AND DISCUSSION

Results of temperature effect study on the resistance of PP filled with 20% wf. carbon black are presented as relative change in electrical resistivity at ρ_T/ρ_{20}, where ρ_T – electrical resistivity at temperature T, ρ_{20} – resistivity at room temperature (Fig. 5.1) this dependence has a clear extremum, and the maximum (peak) electric resistance is in the temperature range of the polymer matrix melting.

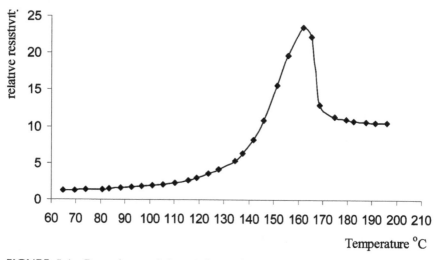

FIGURE 5.1   Dependence of the relative resistivity of PP-20%wf on crystallization temperature.

Represented dependence can be divided into four characteristic temperature range: 1 weak growth resistance with increasing temperature, 2a sharp increase of resistance, and 3a sharp drop of resistance, 4 slight decrease of resistance. Define ranges of these sites allows linearization a of dependencies logarithm of relative resistivity versus the reciprocal of temperature. Throughout the temperature range investigated characteristic curves have a satisfactory linear relationship (R^2 value of reliable approximation from 0.8 to 0.99).

The boundaries of these temperature ranges between the values: (1) range: 60°–120 °C, (2) range: 120°–160 °C; (3) range 160°–171.44 °C, (4) range above 171.44 °C.

The increase of resistance occurs as the temperature rises at the first and second ranges. Temperature coefficient of resistance change is positive for systems in these areas. The activation energy of changes in resistivity when heated PP filled up in the first region is amount to 12 kJ/mol, and the second to 83 kJ/mol. At 3 and 4 ranges of resistance decreases with increasing temperature, the activation energy is as follows: 120 kJ/mol 5.54 kJ/mol, respectively.

The final resistance of the PP CB filler reaches a constant value and is greater than the initial approximately 5 times.

To explain the external dependence of the resistance of PP filled with carbon black it is advisable to raise enough developed apparatus of model representations of heterogeneous processes. The process of changing the conductivity of PP-black with an increase in temperature is associated with changes in the morphology of the system, which largely forms the filler.

Changes in the degree of crystallinity with temperature for unfilled and filled with carbon black PP are shown in Fig. 5.2. The figure shows that the degree of crystallinity of PP increases twice by adding black.

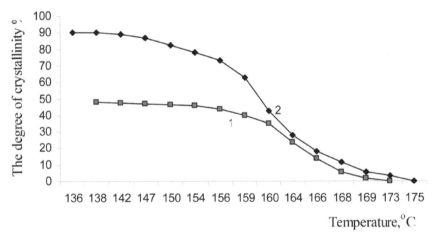

FIGURE 5.2   Dependence of the degree of crystallinity of PP (1) and PP filled 20% wf. carbon black (2) on the temperature.

Formally kinetic analysis of processes in heterophase surfaces can be carried out in terms of topochemical approach, which is carried out on the basis of the theoretical curves (power, exponential) and is based on assumptions about the various laws of the formation and growth of nuclei nucleation [9]. Most often Erofeev equation is used for the analysis of such processes.

$$\alpha = 1 - \exp\left(-kt^n\right) \tag{1}$$

where α is the proportion of reacted material; k is the rate constant; t is the time expansion, n is the kinetic parameter. Value of the exponent n gives an indication of the conditions of formation and growth of nuclei and the reaction mechanism. If the velocity is proportional the weight of the unreacted substance, n is equal 1. When the reaction is located in the kinetic region n > 1. The reaction rate depends only weakly on the nucleation rate and the growth of existing nuclei is determined when n >> 1. If the reaction is limited by diffusion, the greater the deviation from the n units, the greater the effect of diffusion processes (n < 1). Differential form of Eq. (1)

$$\frac{d\alpha}{dt} = nk^{1/n} \left[-\ln(1-\alpha)^{1-1/n} \right](1-\alpha)$$

(2)

From Eq. (2) can obtain an approximate equation:

$$\frac{d\alpha}{dt} = k\alpha^a (1-\alpha)^b$$

(3)

where a and b are constants corresponding to certain values of n [9].

Equation (1) in the form $-\ln(1-\alpha) = kt^n$ (4) is known as the equation Iohansona-Mele Avrami Erofeev Kolmogorov [6] (in the literature it is referred to as the Avrami equation or Kolmogorov Avrami Erofeev). In Eq. (4), k is a generalized rate constant, n is exponent. In Ref. [11] k and n are considered as parameters of the Avrami, depending on the geometry of the growing crystals and the nucleation process. These parameters are also a convenient way to represent empirical data crystallization. The equation describing the crystal growth rate is presented in Ref. [11]

$$v = \frac{d\alpha}{dt} = v_0 e^{-\Delta G/RT}$$

(5)

where α is a degree of crystallinity of the polymer matrix at due time in an isothermal process, ΔG is the total free energy of activation, which is equal to the enthalpy change during crystal growth.

$\Delta G = \Delta G* + \Delta G_\eta$. $\Delta G*$ – the free enthalpy of nucleation of a nucleus of critical size, ΔG_η the free enthalpy of activation of molecular diffusion across the phase boundary. Equating the right hand side of Eqs. (3) and (5), we obtain the expression

$$k\alpha^a (1-\alpha)^b = v_0 e^{-\Delta G/RT}$$

(6)

Studies in Ref. [11], allowed to determine the exponent n Avrami equation Erofeev-Kolmogorov crystallization of PP filled with carbon black. In the initial stage of crystallization according to Ref. [11], it is 2, and the second stage is 3. Isothermal melt crystallization process is reversed so that at low temperatures must take place a decrease of the crystal phase, which was formed on the last (second) stage. Therefore, by solving equation for n = 3 and taking the values of constants for this type of a heterophasic processes a= 2/3, b = 0.7 [9] we obtain the expression

$$2/3\ln\alpha + 0.7\ln(1-\alpha) = \ln\frac{v_o}{k} - \frac{\Delta G}{RT_c} \tag{7}$$

T_c – crystallization temperature for different weight contents of carbon black.

Depending defined for isothermal processes with a certain degree of approximation can be used for processes occurring in the quasi-isothermal conditions (at a relatively slow heating, which is implemented in the research process). Dependence of the degree of crystallinity of PP filled in the considered coordinates is presented in Fig. 5.3.

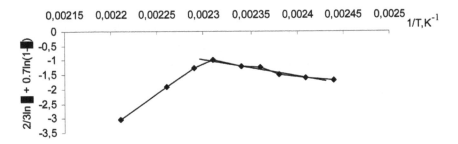

FIGURE 5.3 The crystallinity degree dependence of PP filled 20%wf carbon black on the temperature in the coordinates of the Avrami-Erofeev-Kolmogorov equation.

Dependence in Fig. 5.3 is well approximated by two straight lines intersecting at a temperature of 162 °C. The degree of crystallinity at this temperature is 30%. Analysis of temperature effect on the degree of crystallinity according to Eq. (7) allows to determine ΔG for two characteristic temperature regions, which are determined by the intersection of the lines in Fig. 5.3. Plot for low temperatures (up to 160 °C) ΔG is 47 kJ/mol. The activation energy of the process of increasing resistance in this area is 83 kJ/mol. The difference in the values of defined energies gives reason to assume that the change in conductivity is involved not only a change in the structure of the system, but the result affects some another processes. Is possible these are the fluctuation processes along with the redistribution of the filler particles in increasing the amorphous phase and the agglomerates are destroyed due to the interaction with the polymer segments sites. The activation energy of the process of changing the degree of crystallinity in the area of high temperature (when n = 3) is amount to122 kJ/mol. The activation energy for decreasing the relative resistivity is 120 kJ/mol. Energy values of the two processes are very similar, which gives grounds to consider these as interrelated processes. Thus, not only low-temperature stages, but also in high temperature change affects the degree of crystallinity of the PP-conductivity specification. Since as shown in Ref. [6] molecular segments PP and PP molecules generally have sufficient adsorption energy to the particles carbon black, it can be assumed that the carbon black particles are displaced into the amorphous phase and collected in agglomerates as a result of this change occurs in crystallinity, and electrical resistance at range 1 and 2. However, a certain part of the carbon black particles served as centers of crystallization filled system remains bound to the polymer molecules. Having destroyed such polymer crystals relieve certain portion of the carbon black particles, which either form additional conduction channels or enhance existing ones. As a result, the resistance of the entire system decreases. The results of studies carried out in Ref. [11], allow us to determine the effect of concentration of filler (carbon black) forming the morphology of the polymer matrix composite PP-W (carbon black) and temperature on the characteristic parameter k (Avrami parameter) reflecting the growth of crystals and the process of nucleation (Fig. 5.4).

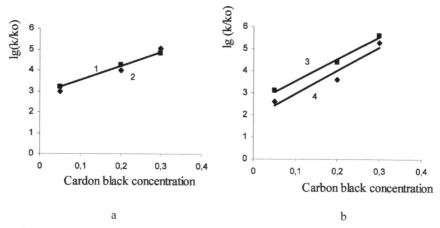

FIGURE 5.4   Dependence of the relative values of the constants in the Eq. (4) on the concentration carbon black at 128C (1) and 129 °C (2) (a) 130 °C (3) and 131 °C (4) (b).

A generalization of explored dependences in the range of investigated concentrations of carbon black and at 128–131 °C gives the dependence of the relative value of the constant k (k/k_o) of the defining parameters in equation

$$\ln (k/k_o) = A(T) \times C + B(T), \tag{8}$$

where C is the concentration of carbon black in polypropylene, A and B – parameters depending on the temperature at which the crystallization process proceeds, k_o – characteristic parameter for pure PP.

Dependence of the parameters A and B have a complex temperature dependence of the third-degree polynomial, which in some approximation can be approximated by a linear dependence with the value of R^2 reliable approximation not less than 0.8. As a result, the values for the Eq. (8).

$$\ln (k/k_o) = (3T\text{-}1191) \times C\text{-}0,7 \times T +265 \tag{9}$$

Comparison of the relative values of the parameters calculated by the proposed correlation and obtained experimentally, is presented in Fig. 5.5. Values coincide satisfactorily, especially 129 and 130 °C. By substituting the value of ln k of Eq. (9) into Eq. (7), we obtain the expression 2/3

$$\ln \alpha + 0.7\ln(1\text{-}\alpha) = \ln v_o - \ln k_o - (3T\text{-}1191)C + 0.7T\text{-}265\text{-}\Delta G/RT \tag{10}$$

or 2/3 ln α+0.7

$$\ln(1\text{-}\alpha) = A\text{-}B(T) - \Delta G/RT \tag{11}$$

where $A = \ln v_o - \ln k_o - 265$, $B(T) = (3T\text{-}1191)C + 0.7T$

Dependence in Fig. 5.6 have an extreme character with a peak at 162 °C and are approximated by two straight lines as in Fig. 5.3. Determined from this dependence activation energy for temperatures up to 160 °C is 48 kJ/mol, which corresponds to the value found from the Eq. (7). For temperatures above 160 °C energy is 373 kJ/mol. Found the energy characteristics of the process changes the crystallinity of PP filled with soot have a value higher than the energy of the relative resistivity change for this system, but the signs of these parameters are the same.

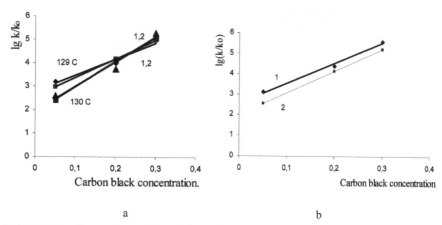

a   b

FIGURE 5.5   Comparison of the relative values of the parameter depending on k, obtained from model calculations (1) and experimental (2) 129 °C and the temperature 130 °C (a) and 131 °C (b).

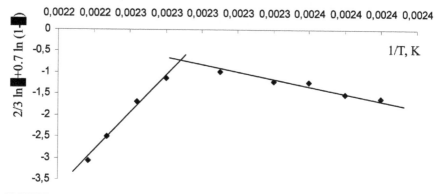

FIGURE 5.6   Dependence of crystallinity degree of carbon black-filled PP on the temperature in the coordinates of Eq. (11).

Dependence of resistivity on the relative degree of crystallinity obtained in the heating speed of 3 deg/min PP, filled with 20%wf carbon black is shown in Fig. 5.7. The dependence has an extreme character: The highest relative resistivity appears at the degree of crystallinity of 30÷60%, which corresponds to a temperature of, 160–162 °C. Depending on the nature of the polynomial describing the dependence $\rho/\rho_{20} = -0.008 \, \alpha^2 + 0.7\alpha + 10.6$.

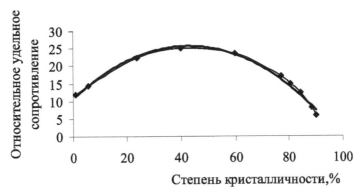

FIGURE 5.7   Relative resistivity dependence on the degree of crystallinity obtained in the process of heating the PP filled with 20% wf. carbon black.

5.4   CONCLUSIONS

During the heating process there is a change in crystallinity and resistance of the composite material based on polypropylene and carbon black. As a result, there are two mechanisms, one of which is realized to 160 °C, and the other is above this temperature. Change the kinetic characteristics both for the crystallization process in the proposed coordinates and for the process of change resistance according to experimental data occurs at the same temperature. For both processes, these dependencies are approximated by linear and have the same sign temperature coefficients. The activation energy of changes in the degree of crystallinity of model representation and the activation energy of resistance change have the same values for the temperature range above 160 °C. It was determined that the resistance value takes the maximum at a temperature of energy change for both processes (at 162 °C). The value of crystallinity degree is equal to 30% under such conditions. The proposed equations to describe the process satisfactorily coincide with the values obtained from the experiment.

KEYWORDS

- **Carbon black**
- **Effect**
- **Filled**
- **Model**
- **Polypropylene**
- **Resistance**
- **Temperature**

REFERENCES

1. Sichel, E. K. (1989). Carbon Black Polymer Composites, New York, Marcel Dekker.
2. Koszkul, J. (1997). Energochemia Ekologia Karbo, *42*, 84.
3. Yamaki, J., Maeda, O., & Katayama, Y. (1975). Electrical Conductivity of Carbon Fiber Composites, Kobunshi Ronbunshi, *32(1)*, 42
4. Moonen, J. T., Vander Putter, D., Brom, H. B. (1991). Physical Properties of Carbon Black, *Polymer Compounds, Synthetic Metals, 41–43*, 969–972.
5. Siohel, E. K., & Basel, N. Y., (1982). Carbon Black Polymer Composites, The Physics of Electrically Conducting Composites, Marcel Dekker, 214c.
6. Horowitz, S., Dirk, L. M. A., Yesselman, J. D., Nimtz, J. S., Adhikari, U., Mehl, R. A., St. Scheiner, R., Houtz, L., Al-Hashimi, H. M., & Trievel, R. C. (2013). Conservation and Functional Importance of Carbon-Oxygen Hydrogen Bonding in Ado Met-Dependent Methyl transferases, J. Am. Chem. Soc., 135, 15536–15548.
7. Shevchenko, V. G., Ponomarenko, A. T., Carl Klason, Tchmutin, I. A., & Ryvkina, N. G. (1997). Electromagnetic Properties of Synthetic Dielectrics from Insulator-Coated Conducting Fibers in Polymeric Matrix, *J. Electromagnetics, 17(2)*, 157–170.
8. Jauffres, D., Lame, O., Vigier, G., & Dore, F. (2007). Micro Structural Origin of Physical and Mechanical Properties of Ultra High Molecular Weight Polyethylene Processed by High Velocity Compaction, *Polymer, 48(21)*, 6374–6383.
9. Mukherjee, P., Drew, M. G. B., Gómez-Garcia, C. J., & Ghosh, A. (2009). (Ni2), (Ni3), and (Ni2 + Ni3) a Unique Example of Isolated and Cocrystallized Ni2 and Ni3 Complexes, *Inorg. Chem., 48*, 4817–4825.
10. Nelson, J. H., Howels, P. N., Landen, G. L., De Lullo, G. S., & Henry, R. A. (1979). Catalytic Addition of Electrophiles to β-dicarbonyles, In Fundamental Research in Homogeneous Catalysis, 3, (New York, London Plenum), 921–939.
11. Mucha, M., Marszalek, J., & Fidrych, A. (2000). Crystallization of Isotactic Polypropylenes Containing Carbon Black as Filler, *Polymer, 41*, 4137–4142.

CHAPTER 6

A RESEARCH NOTE ON KINETICS OF RELEASE MEDICINAL SUBSTANCES FROM CHITOSAN FILMS

ANGELA SHURSHINA, ELENA KULISH, and ROMAN LAZDIN

CONTENTS

ABSTRACT

The kinetics of a controlled release of medicinal substances from films on the basis of polymer chitosan is investigated and diffusion coefficients are calculated. The kinetic curve releases of antibiotics having the abnormal character are shown. The analysis of the obtained data showed that the weak chemical modification is the possible direction of monitoring of transport properties medicinal the chitosan films.

6.1 INTRODUCTION

Recently the considerable interest to studying of the diffusion phenomena in polymers is observed. The reason is that the diffusion plays a primary role in such processes as dialysis, permeability of bio membranes and tissues, prognostication of the protective properties of polymeric coatings, etc. [1]. Studies of the diffusion of water into a polymer matrix are also necessary in development of medicinal polymeric films with controllable release of a medicinal preparation.

Research of regularities of processes of diffusion of medicinal substance in polymeric films and opportunities of monitoring of a release of medicinal preparations became the purpose of this work. As a matrix for an immobilization of medicinal preparations is used the polysaccharide of a natural origin chitosan, as medicine antibiotics Gentamicin and Cefatoxim is used.

6.2 EXPERIMENTAL PART

We used a sample of chitosan (ChT) manufactured at Bioprogress ZAO (Russia) by alkaline deacetylation of crab chitin (degree of deacetylation ~84%) with $M_{sd} = 334,000$. As the medicinal substance (MS) served antibiotics of the amino glycoside (gentamicin, GM) and cefalosporin series (cefatoxim, CFT).

Film samples of chitosan acetate (ChTA) were produced by casting of a polymer solution in 1% acetic acid onto the glass surface. The films were dried in a vacuum box at a residual pressure of 10–15 mm Hg. An aqueous solution of an antibiotic was added to the ChT solution immediately before the fabrication of the films. The contents of a MS in a fi lm were 0.01, 0.05, and 0.1 mol mol^{-1} ChT. The fi lm thicknesses were maintained constant (100 μm) in all the experiments. To preclude their dissolution in water, the films samples were isothermally annealed at a temperature of 120 °C.

To study the release kinetics of the drug, the sample was placed in a cell with distilled water. The antibiotic which was released into the aqueous phase was registered by spectrophotometry at the wave length corresponded to the drug release maximum in the UV-spectrum. The drug quantity released from the film at the moment t (G_s) was estimated by the calibration curve. The moment of steady drug concentration in the solution (G_∞) was treated as equilibrium. The drug mass fraction α available for diffusion was estimated as the relation of the maximum quality of the antibiotic released from the film and its quantity introduced into it.

The interaction of MS with ChT was studied by IR and UV spectroscopies. The IR spectra of the samples were recorded with a Shimadzu spectrometer (KBr pellets, films) in the spectral range 700–3600 cm^{-1}. UV spectra of all the samples were recorded in 1-cm-thick quartz cuvettes relative to water with a Specord M-40 spectrophotometer in the range 220–350 nm.

To determine the mass fraction of the MS chemically bound to the polymer matrix (β), adducts formed in the interaction between the ChT and the antibiotic in an acetic acid solution were isolated by double reprecipitation into a NaOH solution, followed by washing of the precipitate formed with ethanol. The precipitate was dried to constant weight. The amount of MS bound to the chitosan matrix was determined by elemental analysis with a EUKO EA-3000 analyzer and by UV spectrophotometry.

6.3   RESULTS AND DISCUSSION

Now the established fact is standards that at a controlled release of MS from polymeric systems diffusion processes dominate.

In Fig. 6.1, typical experimental curves of an exit of MS from chitosan films with different contents of MS are presented. All the kinetic curves are located on obviously expressed limit corresponding to an equilibrium exit of MS (G_Ψ).

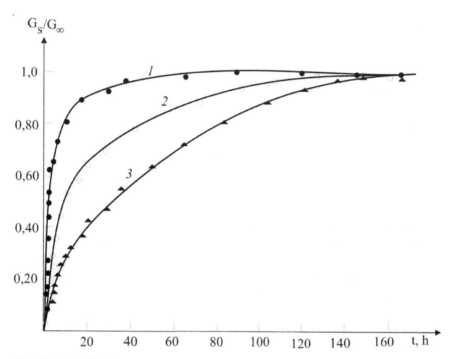

FIGURE 6.1   Kinetic curves of the release of the MS from film systems ChT-CFT with the molar ratio of 1:0.01 (1), 1:0.05 (2) and 1:0.1 (3) Isothermal annealing times 30 min.

Mathematical description of the desorption of low molecular weight component from the plate (film) with a constant diffusion coefficient is in detail considered in Ref. [2], where the original differential equation adopted formulation of the second Fick's law:

$$\frac{\partial \tilde{n}_s}{\partial t} = D_s \frac{\partial^2 c_s}{\partial x^2} \tag{1}$$

The solution of this equation disintegrates to two cases: for big $(G_s/G_\infty > 0.5)$ and small $(G_s/G_\infty \leq 0.5)$ experiment times:

$$\text{On condition } G_s/G_\infty \leq 0.5 \tag{2a}$$

$$\text{On condition } G_s/G_\infty > 0.5 \tag{2b}$$

where $G_s(t)$—concentration of the desorbed substance at time t and G_∞ -Gs value at $t \to \infty$, L – thickness of the film sample.

In case of transfer MS with constant diffusion coefficient the values calculated as on initial (condition $G_s/G_\infty \leq 0.5$), and at the final stage of diffusion (condition $G_s/G_\infty > 0.5$), must be equal. The equality $D_s^a = D_s^b$ indicates the absence of any complications in the diffusion system polymer low-molecular substance [3]. However, apparently from the data presented in Table 6.1, for all analyzed cases, value of diffusion coefficients calculated at an initial and final stage of diffusion don't coincide.

TABLE 6.1   Parameters of Desorption of Cefatoxim, Gentamicin from Chitosan Films

Film composition	MS content of a film, mol/mol⁻¹ ChT	Annealing duration, min	$D_s^a \times 10^{11}$, cm²/sec	$D_s^6 \times 10^{11}$, cm²/sec	n	α
ChT-GM	1:0.01	30	25.30	1.99	0.31	0.97
		60	23.70	1.92	0.24	0.92
		120	22.00	1.50	0.21	0.88
	1:0.05	30	24.60	1.87	0.20	0.84
	1:0.1	30	23.50	1.59	0.16	0.73
ChT-CFT	1:0.01	30	91.40	6.46	0.38	0.97
		60	83.90	5.43	0.36	0.95
		120	77.60	5.18	0.31	0.94
	1:0.05	30	56.80	6.13	0.32	0.94
	1:0.1	30	40.90	5.79	0.28	0.92

This indicates to a deviation of the diffusion of the classical type and to suggest the so-called pseudonormal mechanism of diffusion of MS from a chitosan matrix.

About pseudo normal type diffusion MS also shows kinetic curves, constructed in coordinates $G_s/G_\infty - t^{1/2}$ (Fig. 6.2). In the case of simple diffusion, the dependence of the release of the MS from film samples in coordinates $G_s/G_\infty - t^{1/2}$ would have to be straightened at all times of experiment. However, as can be seen from (Fig. 6.2), a linear plot is observed only in the region $G_s/G_\infty < 0.5$, after which the rate of release of the antibiotic significantly decreases.

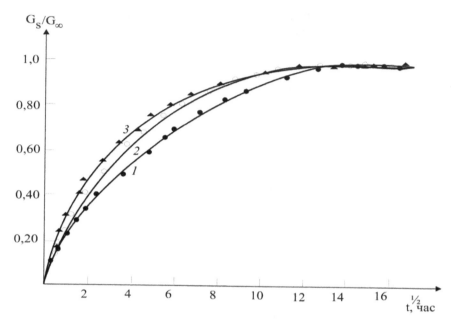

FIGURE 6.2   Kinetic curves of the release of the MS from film systems ChT-GM with a molar concentration of 0.01 (1) 0.05 (2) and 0.1 (3) Isothermal annealing time 30 min.

At diffusion of MS from films take place anomalously low values of the parameter n estimated on a tangent of angle of an inclination in coordinates of lg (G_s/G_∞)–lgt. The increase in concentration of MSV and time of isothermal annealing is accompanied by padding decrease of parameter n. Simbatno to an index of n changes also the sizea.

All the specific features inherent in the anomalous (non-Fick) diffusion are well described in terms of the relaxation model [4]. In contrast to the Fick diffusion, in which an instantaneous attainment of the surface concentration of the sorbate and its change is assumed, the relaxation model presumes variation of the concentration in the surface layer (i.e., variable boundary conditions) in accordance with the first-order equation [5]. One of the main reasons causing change of boundary conditions, call a nonequilibrium of the structural-morphological organization of a polymeric matrix [4]. It is noteworthy that the anomalously small values of n have also been observed by other researchers and attributed to the strong interaction of the MS with the polymer [6].

The structure of the medicinal substances used in the study suggest that they interact with ChT to give, for example, ChT-antibiotic complexes via hydrogen bonds and (or) polymeric salts formed as a result of an exchange interaction. The interaction of antibiotics with ChT is evidenced by UV spectroscopic data. The maximum absorption of CFT at its concentration of 10^{-5} M in 1% acetic acid is observed at 261 nm. Upon addition of an equivalent amount of ChT to the solution, the intensity of the absorption peak of the medicinal preparation noticeably grows and its position is bathochromically shifted by approximately 5–10 nm. The UV spectrum of GM at a concentration of 10^{-2} M in 1% acetic acid shows an absorption peak at 286 nm. Addition of a ChT solution to the GM solution results in that a precipitate is formed; however, analysis of the supernatant fluid shows that the peak of the corresponding absorption band in its spectrum is shifted by 5–7 nm. The observed changes unambiguously indicate that ChT affects the electron system of MS and adducts are formed. The binding energies in the complexes, evaluated by the shift of the absorption peaks in the UV spectra, are about 10 kJ mol^{-1}. This suggests that the Complexation occurs via hydrogen bonds.

The interaction of the antibiotics under study with ChT is also confirmed by IR spectroscopic data. For example, the IR spectrum of ChT shows absorption bands at 1640 and 1560 cm^{-1}, associated with deformation vibrations of the acetamide and amino groups, and at 1458 and 1210 cm^{-1}, associated with planar deformation vibrations of hydroxy groups. Analysis of the IR spectra of adducts formed by interaction of ChT with antibiotics suggests that some changes occur in the IR spectra. For example, absorption bands associated with stretching vibrations of the C=O and C=N groups of the medicinal compound appear at 1750 and 1710 cm^{-1} in the IR spectra of ChT-CFT polymeric adducts, and a strong absorption band related to the SO$_4^{2-}$ group appears at 619 cm^{-1} in the IR spectrum of the ChT-GM reaction product. In addition, the IR spectra of all the compounds being analyzed show that the absorption band at 3000–3500 cm^{-1}, associated with stretching vibrations of OH and NH groups, is broadened as compared with the corresponding bands of the antibiotics and ChT, which, taken together, suggests that ChT–antibiotic complex compounds are formed via hydrogen bonds.

Of no lesser importance for interpretation of the diffusion data are exchange interaction between ChT and MS, especially for GM. Because the sulfate-anions are dibasic, it can be assumed that two kinds of salts are formed, which provide cross-linking of ChT molecules, with loss of solubility: (i) a water-insoluble "double" salt, ChT–GM sulfate, and (ii) a mixture of salts, water-insoluble ChT sulfate and soluble GM acetate.

In the case of CFT, the exchange reaction between ChT acetate and CFT yields soluble salts. Accordingly, the ChT–MS reaction product will be then composed of only the H-complex of ChT–CFT.

The data on the fraction β of the antibiotic bound into polymeric adducts formed in acetic acid solutions are presented in Table 6.2.

TABLE 6.2   Mass Fraction β of An Antibiotic in the Reaction Adducts p Obtained from 1% Acetic Acid

Antibiotic	MS concentration in a film, mol mol^{-1} CTS	β
GM	1.00	0.69
	0.10	0.30
	0.05	0.20
	0.01	0.07
CFT	1.00	0.16
	0.10	0.09
	0.05	0.05
	0.01	0.02

It can be seen that the fact that GM can "cross-link" chitosan chains results in that in the substantially larger amount of the MS firmly bound macromolecules, compared with CFT.

Thus, formation of chemical compounds of MS with ChT probably is the reason of being observed anomalies decreases of rate of release of MS from a film owing to diffusion, and also decrease of a share of an allocated medicinal (α). Mild chemical modification, for example by cross-linking macromolecules salt formation, not affecting the chemical structure of the drug, is a possible area of control of the transport properties of medicinal chitosan films.

KEYWORDS

- **Chitosan**
- **Diffusion**
- **Medicinal substance**
- **The abnormal character**

REFERENCES

1. Iordansky, A. L., Shterezon, A. L., Moiseev, Yu. B. & Zaikov, G. E. (1979). Uspehi Himii, XLVIII, *8,* 1460.
2. Crank, J. (1975). The Mathematics of Diffusion, Clarendon Press, Oxford.
3. Livshic, V. A., Bonarcev, A. P., Iordansky, A. L. et al. (2009). Vysokomol Soedin, *51(7),* 1251.
4. Malkin, A. Yu., & Tchalykh, A. E. (1979). Diffuziya i vyazkost polimerov Metodyi izmereniya Himiya, Moscow.
5. Pomerantsev, A. L. (2003). Dis Metodyi Nelineinogo Regressionnogo Analiza Dlya Modelirovaniya Kinetiki Himicheskih I Fizicheskih Processov, D-Ra Fiz-Mat Nauk. M, MSU.
6. Smotrina, T. V. (2012). Butlerov Soobshch, *29(2),* 101.

CHAPTER 7

VISCOMETRY STUDY OF CHITOSAN IN ACETIC ACID SOLUTION: A RESEARCH NOTE

VALENTINA CHERNOVA, IRINA TUKTAROVA, ELENA KULISH, GENNADY ZAIKOV, and ALFIYA GALINA

CONTENTS

ABSTRACT

Some ways of estimating the values of the intrinsic viscosity of chitosan were analyzed. It is shown that the method of estimating the current value of the intrinsic viscosity of Irzhak and Baranov can adequately assess the conformational state of the macromolecular coil and its degree of swelling.

7.1   INTRODUCTION

Recently there have been a large number of studies related to the producing and studying of chitosan-based film materials for using them in different areas [1]. According to modern views, the supra molecular structure of the polymer films formed from the solution is determined by the conformational state of the polymer in the initial solution, which in turn, predetermines the complex physicochemical properties of the material. Viscosimetry is one of the most accessible and informative methods of studying the conformational state of polymers in solution. However, in the case of using viscosimetry to study polyelectrolyte solutions, reliable determination of intrinsic viscosity to assess their conformational state, faces with certain difficulties. In this paper we attempted to analyze some of the ways to estimate the values of the intrinsic viscosity of chitosan, which when dissolved in acidic aqueous media (e.g., acetic acid) acquires the properties of polyelectrolyte.

7.2   EXPERIMENTAL PART

As objects of study used a sample of chitosan (manufactured by "Bioprogress" Schyolkovo) obtained by alkaline deacetylation of crab chitin (deacetylation degree of ~84%). The molecular weight of the original chitosan (CHT) was determined by combination of sedimentation velocity method and viscosimetry according to the formula [2]:

$$M_{s\eta} = \left(\frac{S_0 \eta_0 [\eta]^{1/3} N_A}{A_{hi}(1 - \nu\rho 0)} \right)^{3/2}$$

Where is S_0 – sedimentation constant; η_0 – dynamic viscosity of the solvent, which is equal to 1.2269×10^{-2} PP; $[\eta]$ – intrinsic viscosity, dl/g; N_A – Avogadro's number, equal to 6.023×10^{23} mol^{-1}; $(1-\nu\rho_0)$ –Archimedes factor or buoyancy factor, ν – the partial specific volume, cm^3/g, ρ_0–density of the solvent g/cm^3; A_{hi}– hydrodynamic invariant equal to 2.71×10^6.

The experiments were carried out in a two-sector 12 mm aluminum cell at 25 °C. Thermo stating was carried out with an accuracy of 0.1 °C. Sedimentation constant was calculated by measuring the boundary position and its offset in time by means of optical schemes according to the formula [2]:

$$S_0 = \frac{1}{\omega^2 X_{max}} \frac{dX_{max}}{dt}$$

Archimedes factor was determined psychometrically by standard methods [2] and was calculated using the following formulas:

$$v = v_0 \left[\frac{100}{m_0 \rho} \left(\frac{1}{m_0} - \frac{1}{m} \right) \right]$$

where is v_0 – volume of the pycnometer; m_0 – weight of the solvent in the pycnometer; m – weight of the solution in the pycnometer; ρ=100 g /m_0; (g – solution concentration in g/mL). M_{sh} value for the CHT sample used in was 113,000 Da.

In order to determine the constants K and α in Mark-Houwink-Kuhn equation the sample of initial chitosan was fractionated to 10 fractions, for each of them has been determined $M_{s\eta}$ value and an intrinsic viscosity value. Fractionation of chitosan was performed by sequential deposition [3]. As the precipitant was used acetone. Viscosimetry studies were performed according to standard procedure on an Ubbelohde viscometer at 25 °C [4].

As the solvent was used acetic acid at a concentration of 1, 10 and 70 g/dL. Value of intrinsic viscosity of the initial CHT sample and its fractions in solutions of acetic acid was determined in two ways by the method of Fuoss [5] and the method of Irzhak and Baranov [6, 7]. Dilution of initial solution of the polymer in the calculation of intrinsic viscosity by Irzhak and Baranov carried out by the solvent, while using the method Fuossa by water.

7.3   RESULTS AND DISCUSSION

Most often the values for determining the intrinsic viscosity [η] of Polymer solutions are described by Huggins equation [8]:

$$\eta\_sp / c = [\eta] + k\_1[``\eta``] ^2 c \tag{1}$$

where is c – concentration of the polymer in the solution (g/dL); k_1 – constant reflecting the reaction of the polymer with a solvent; η_{sp} – specific viscosity of polymer solution equal to: $\eta_{sp}=\eta_{rel}-1$. Relative viscosity η_{rel} is the ratio viscosity of the polymer solution (η_p) to the viscosity of the solvent (η_0). The main condition for the applicability of the Huggins equation is linear dependence of the reduced viscosity (η_{sp}/c) of the polymer concentration in the solution and the absence of strong intermolecular interactions between the components macro chains. Value of intrinsic viscosity is calculated by extrapolation to zero concentration dependence of the reduced viscosity of the polymer concentration in the solution.

However, in the case of studies of solutions of poly electrolytes in dilute solutions the macromolecular coil swells considerably, because there is a well-known effect of polyelectrolyte swelling. As a consequence there is a deviation from the linear dependence of the viscosity of the polymer concentration in the solution [9, 10]. For example, when studying solution polyelectrolyte HTZ in acetic acid, the experimental data in the viscosity of dilute solutions are also not described by the Huggins equation.

Fuoss and Strauss [8] proposed an empirical equation to determine the intrinsic viscosity of poly electrolytes:

$$\frac{\eta_{sp}}{c} = \frac{[\eta]}{1 + B\sqrt{c}} \tag{2}$$

where is B–coefficient characterizing the electrostatic interaction of the poly ion with simple ions.

When using Fuoss and Strauss approach dilution of the original concentration of the polymer solution is made by water instead of solvent. In the case where the solvent is a strong electrolyte (e.g., hydrochloric acid), the dilution water does not change the degree of dissociation of the solvent and does not change the ionic strength of the solution. Hence the value of the intrinsic viscosity, determined by extrapolation to zero concentration will indeed reflect the size of the coil in the solvent being. In our case, acetic acid is used as the solvent, which is weak and which dilution water is accompanied by an increase in its degree of dissociation. Increasing the degree of dissociation of acetic acid leads to additional protonation CHT macromolecules in solution and causes unfolding of the macromolecular coil. As a result, the value of intrinsic viscosity of CHT determined by Fuoss equation will not reflect the conformation of macromolecules in a solution of acetic acid concentrations investigated, but will reflect CHT conformation in solution infinitely dilute acetic acid.

More reasonable method for determining the intrinsic viscosity of polyelectrolytes in order to assess conformational state of the coil is considered the reception of isoionic dilution [11]. At the same time, dilution of the polyelectrolyte solution is made by low-molecular salt. This method is very time consuming, because it requires optimization. Meanwhile, in Ref. [9, 10], it was suggested that adequate alternative to method isoionic dilution is the method of determining the intrinsic viscosity of Irzhak and Baranov (method of assessing the current intrinsic viscosity), by the equation:

$$[\eta] = \frac{[\ln]}{c} \qquad (3)$$

Using Eq. (3) the intrinsic viscosity can be determined by the initial slope of the $\ln \eta_{rel}$ from c. In this case, we can estimate $[\eta]$ of polyelectrolyte excluding inevitable, in our case, "overstatement" values $[\eta]$, concomitant determination of the intrinsic viscosity by the Eq. (2). Comparison of values of the intrinsic viscosity of the initial sample CHT defined by both analyzed methods (Fuoss and Irzhak-Baranov methods) in acetic acid solution of 1, 10 and 70% indicates that the value $[\eta]$, defined by Eq. (3) in all cases less than values defined by Eq. (2) (Table 7.1).

TABLE 7.1 Some Characteristics of Chitosan Calculated Using Methods Fuoss and Irzhak-Baranov

Concentration of acetic acid, %	Fuoss method				Irzhak-Baranov method			
	α	$K \times 10^5$	$[\eta]$, g/dL	M^*, kDa	α	$K \times 10^5$	$[\eta]$, g/dL	M^*, kDa
1	1.15	2.89	12.20	77.9	1.02	5.53	7.79	111.6
10	1.03	10.11	11.30	79.7	0.93	13.91	6.89	113.5
70	0.90	37.41	10.09	83.8	0.81	41.39	5.25	116.3

* M calculated by the Mark-Houwink-Kuhn equation.

After fractionation of the initial sample of CHT and determining the molecular weight of each fraction the values of intrinsic viscosity of fractions in 1, 10, and 70% acetic acid were determined by two analyzed methods by Fuoss and by Irzhak and Baranov. This made it possible to estimate the parameters K and α in Mark-Houwink-Kuhn equation, testifying about the conformational state of macromolecules in CHT solution. Comparison of the values α in Mark-Houwink-Kuhn equation defined by two analyzed methods shows that for all the studied concentrations of acetic acid the macromolecular coil "by Fuoss" has more comprehensive conformation than the coil "by Irzhak." The reason for the discrepancy in the values of K and α, defined using two methods of determining the viscosity obviously is related to the fact that when using the Fuoss method the coefficients in the Mark-Kuhn-Houwink equation doesn't reflect CHT conformation in solution of acetic acid concentrations studied, but reflects CHT conformation in a dilute solution of acetic acid.

7.4. CONCLUSIONS

Value of the molecular weight calculated from the values $[\eta]$, α and K according to the Irzhak and Baranov equation satisfactorily coincide CHT molecular weight obtained by the absolute method in acetate buffer, $M_{s\eta} = 113,000$ Da. Thus, in the study of polyelectrolyte solutions in weak electrolytes, when calculating the values of the intrinsic viscosity is more correct to use the method of Irzhak and Baranov, rather than the more common method of Fuoss.

KEYWORDS

- **Chitosan**
- **Conformation**
- **Polyelectrolytes**
- **Viscosimetry**

REFERENCES

1. Vikhoreva, G. A., & Galbraikh, L. S. (2002). Plenki i Volokna Na Osnove Hitozana [in] Skryabin, K. G., Vikhoreva, G. A., Varlamov, V. P. et al. Hitin i Hitozan Poluchenie, Svoistva i Primenenie, Nauka, Moscow, 254–279.
2. Rafikov, S. R., Budtov, V. P., Monakov, Yu. B. (1978). Vvedenie v Fiziko-Khimiyu Rastvorov Polimerov Khimiya, Moscow.
3. Nudga, L. A., Plisko, E. A., & Danilov, S. N. (1971). *Journal Obshei Himii, 41,* 2555.
4. Tverdokhlebova, I. I. (1981). Konformatziya Macromolecule (Viskozimetricheskiy Method Otzenki), Khimiya, Moscow.
5. Fuoss, R. M. (1948). Viscosity Function of Poly Electrolytes, *Journal of Polymer Science, 3.* 603.
6. Baranov, V. G., Brestkin, Yu. V, Agranova, S. A. et al. (1986). Vysokomoleculyarnie Soedineniya, *28B,* 841.
7. Pavlov, G. M., Gubarev, A. S., Zaitseva, I. I. et al. (2007). *Journal Prikladnoy khimii, 79,* 1423.
8. Huggins, M. L. (1942). *Journal of the American Chemical Society, 64,* 2716.
9. Tsvetkov, V. N., Eskin, V. E., Frenkel, C. Ya. (1964). Struktura Macromolecule v Rastvorah. Nauka, Moscow.
10. Tiger, A. A. (2007). Fiziko-Khimia Polymerov Nuchniy mir, Moscow.
11. Tanford, Ch. (1964). Fizicheskaya Khimiya Polimerov Khimiya, Moscow.

CHAPTER 8

THE ROLE OF H-BONDING INTERACTIONS (AND SUPRAMOLECULAR NANOSTRUCTURES FORMATION) IN THE MECHANISMS OF HOMOGENOUS AND ENZYMATIC CATALYSIS BY IRON (NICKEL) COMPLEXES

L. I. MATIENKO, L. A. MOSOLOVA, V. I. BINYUKOV, E. M. MIL, and G. E. ZAIKOV

CONTENTS

ABSTRACT

The role played by H-bonds in the mechanisms of homogeneous catalysis is discussed. A strategy of controlling the catalytic activity of $Fe^{II, III}(acac)_n \times L^2$ complexes (L^2 = R_4NBr or 18-crown-6) by adding H_2O to small concentrations is proposed. The activity of Fe catalysts in micro steps of chain initiation and chain propagation in the radical chain process of ethyl benzene oxidation is assessed. The AFM method has been used for research of possibility of the stable supra molecular nanostructures formation based on Dioxygenases models: iron complexes $Fe^{III}_x(acac)_y 18C6_m(H_2O)_n$, $Fe_x(acac)_y CTAB_m$ and nickel complexes $Ni_{xL}{}^1{}_y(L^1{}_{ox})_{z(L}{}^2)_n(H_2O)_m$, with the assistance of intermolecular H-bonds.

8.1   INTRODUCTION

In recent years, the studies in the field of homogeneous catalytic oxidation of hydrocarbons with molecular oxygen were developed in two directions, namely, the free-radical chain oxidation catalyzed by transition metal complexes and the catalysis by metal complexes that mimic enzymes. Low yields of oxidation products in relation to the consumed hydrocarbon (RH) caused by the fast catalyst deactivation are the main obstacle to the use of the majority of biomimetic systems on the industrial scale [1, 2].

However, the findings on the mechanism of action of enzymes, and, in particular, Dioxygenases and their models, are very useful in the treatment of the mechanism of catalysis by metal complexes in the processes of oxidation of hydrocarbons with molecular oxygen. Moreover, as one will see below, the investigation of the mechanism of catalysis by metal complexes can give the necessary material for the study of the mechanism of action of enzymes.

The problem of selective oxidation of alkylarens to hydro peroxides is economically sound. Hydro peroxides are used as intermediates in the large-scale production of important monomers. For instance, propylene oxide and styrene are synthesized from a-phenyl ethyl hydro peroxide, and cumyl hydro peroxide is the precursor in the synthesis of phenol and acetone [1, 2]. The method of modifying the Ni^{II} and $Fe^{II,III}$ complexes used in the selective oxidation of alkylarens (ethyl benzene and cumene) with molecular oxygen to afford the corresponding hydro peroxides aimed at increasing their selectivity's has been first proposed by Matienko [1, 2]. This method consists of introducing additional mono- or multi dentate modifying ligands L^2 into catalytic metal complexes $ML^1{}_n$ ($M = Ni^{II}$, $Fe^{II, III}$, L2 = acac⁻, enamac⁻). The mechanism of action of such modifying ligands was elucidated. New efficient catalysts of selective oxidation of ethyl benzene to a-phenyl ethyl hydroperoxide (PEH) were developed [1, 2]. Role of H-Bonding in mechanism of catalysis with active catalytic particles was established [2]. The high stability of effective catalytic complexes, which formed in the process of catalytic selective oxidation of ethyl benzene to PEH, seems to be associated with the formation of supra molecular structures due to intermolecular H-bonds.

8.2 THE ROLE OF HYDROGEN-BONDS IN MECHANISMS OF HOMOGENEOUS CATALYSIS

As a rule, in the quest for axial modifying ligands L^2 that control the activity and selectivity of homogeneous metal complex catalysts, attention of scientists is focused on their steric and electronic properties. The interactions of ligands L^2 with L^1 taking place in the outer coordination sphere are less studied; the same applies to the role of hydrogen bonds, which are usually difficult to control [3, 4].

Secondary interactions (hydrogen bonding, proton transfer) play an important role in the dioxygen activation and it's binding to the active sites of metallo enzymes [5]. For example, respiration becomes impossible when the fragments responsible for the formation of H-bonds with the Fe-O$_2$ metal site are removed from the hemoglobin active site [6]. Moreover, the O$_2$-affinity of hemoglobin active sites is in a definite relationship with the network of H-bonds surrounding the Fe ion. The dysfunction of Cytochrome P450 observed upon the cleavage of its H-bonds formed with the Fe-O2 fragment demonstrated the important role of H-bonds that form the second coordination sphere around metal ions of many proteins [7].

In designing catalytic systems that mimic the enzymatic activity, special attention should be paid to the formation of H-bonds in the second coordination sphere of a metal ion.

Transition metal β-diketonates are involved in various substitution reactions. Methine protons of chelate rings in β-diketonate complexes can be substituted by different electrophiles (E) (formally, these reactions are analogous to the Michael addition reactions) [8–10]. This is a metal-controlled process of the C−C bond formation [10]. The complex $Ni^{II}(acac)_2$ is the most efficient catalyst of such reactions. The rate-determining step of these reactions is the formation of a resonance-stabilized zwitterion $[(M^{II}L^1_n)^{+E-}]$ in which the proton transfer precedes the formation of reaction products [1, 2]. The appearance of new absorption bands in electron absorption spectra of $\{Ni(acac)_2+L^2+E\}$ mixtures, that can be ascribed to the charge transfer from electron-donating ligands of complexes $L^2 \cdot Ni(acac)_2$ to π-acceptors E (E is tetra cyan ethylene or chloranil) supports the formation of a charge-transfer complex $L^2 \cdot Ni(acac)_2 \cdot E$ [1, 2]. The outer-sphere reaction of the electrophile addition to γ-C in an acetyl acetonate ligand follows the formation of the charge-transfer complex.

In our works we have modeled efficient catalytic systems $\{ML^1n + L^2\}$ (M=Ni, Fe, L^2 are crown ethers or quaternary ammonium salts) for ethyl benzene oxidation to a-phenyl ethyl hydro peroxide, that was based on the established (for Ni complexes) and hypothetical (for Fe complexes) mechanisms of formation of catalytically active species and their operation [1, 2]. Selectivity $(S_{PEH})_{max}$, conversion, and yield of PEH in ethyl benzene oxidation catalyzed by these systems were substantially higher than those observed with conventional catalysts of ethyl benzene oxidation to PEH [1, 2].

The high activity of systems $\{ML^1n + L^2\}$ (L^2 are crown ethers or quaternary ammonium salts) is associated with the fact that during the ethyl benzene oxidation, the active primary $(M^{II}L^1_2)_x(L^2)_y$ complexes and heteroligand $M^{II}_{xL^1{}_y}(L^1{}_{ox})_{z(L^2)}{}^2)_n$ complexes are formed to be involved in the oxidation process.

We established, that in complexes $ML^1_2 \times L^2$, the axially coordinated electron-donating ligand L^2 controls the formation of primary active complexes and the subsequent reactions of β-diketonate ligands in the outer coordination sphere of these complexes. The coordination of an electron-donating extraligand L^2 with an $M^{II}L^1_2$ complex favorable for stabilization of the transient zwitterion $L^2[L^1M(L^1)^{+O_2-}]$ enhances the probability of region selective O_2 addition to the methine C–H bond of an acetyl acetonate ligand activated by its coordination with metal ions. The outer-sphere reaction of O_2 incorporation into the chelate ring depends on the nature of the metal and the modifying ligand L^2 [1, 2]. Thus for nickel complexes, the reaction of acac-ligand oxygenation follows a mechanism analogous to those of Ni^{II}-containing Acireductone Dioxygenase (ARD) [11] or Cu- and Fe-containing Quercetin 2, 3-Dioxygenases [12, 13]. Namely, incorporation of O_2 into the chelate acac-ring was accompanied by the proton transfer and the redistribution of bonds in the transition complex leading to the scission of the cyclic system to form a chelate ligand OAc⁻, acetaldehyde and CO (in the Criegee rearrangement, Scheme 8.1).

In the effect of iron (II) acetyl acetonate complexes, one can find an analogy with the action of Fe^{II}-ARD [11] or Fe^{II}-acetyl acetone Dioxygenase (Dke1) (Scheme 8.2) [14].

In recent years, attention was focused on the interaction of enzyme molecules with water molecules, which is of critical importance for the enzymatic activity [15, 16]. Water molecules present in a protein active site may play not only the structuring role (as a nucleophile and proton donor) but also serve as a reagent in biochemical processes.

Thus the first stage in the O–O bond cleavage in an iron-coordinated H_2O_2 molecule (in horseradish per oxidase) is the proton transfer to the histidine residue (His42), which is made easier when a water molecule is present in the per oxidase active site [17]. As was demonstrated by AB initio a calculation, in the absence of a water molecule, the energy barrier of this reaction is much higher (20 kcal mol^{-1}) as compared with its value found experimentally (1.5±0.7 kcal mol^{-1}).

The role of hydrogen bonds in the mechanism of action of haemoxygenase (HO), metal-free enzyme, in which water acts as a ligand was studied [18]. It was shown that in the absence of H-bonds, water axially coordinated with haem Fe has a weak effect on the nearest amino-acid environment of the haem. With the formation of a system of H bonds involving both free and coordinated water molecules, the situation fundamentally changes. The effect of hydro peroxides on the haem oxygenase-mediated haem (iron proto-porphyrin IX) decomposition to biliverdin, iron and CO was studied [19]. Haem oxygenase uses haem as the substrate and the cofactor. Apparently, in the first stage, meso-hydroxylation occurs, which is followed by its conversion into verdoheme and, ultimately, the opening of the verdoheme ring (under the action of ROOH or O_2) to form biliverdin.

To get insight into the true mechanism of certain organic reactions catalyzed by transition metals, it is necessary to take into account the presence of water [20]. Thus water is a nucleophile in the palladium-catalyzed oxidative carbo hydroxylation of allene-substituted conjugated dienes [21]. This is an example of palladium-mediated diene oxidation accompanied by the formation of a new C–C bond. The reaction proceeds via a (π-allyl) palladium intermediate, which is attacked by the water molecule on the allylic carbon atom. Insofar as the changes in the water concentration exerted different effects on the intra and extradiol oxygenation of 3,5-di(tert-butyl) catechol with dioxygen in aqueous tetra hydrofuran catalyzed by $FeCl_2$ or $FeCl_3$, it was assumed that different intermediates were formed in these reactions. The extradiol oxygenation proceeded selectively only with Fe^{2+} salts, in contrast to Fe^{3+} salts (a model of Catechol 2, 3-Dioxygenase) [22].

At present, only few examples are known where small concentrations of water ($\sim 10^{-3}$ mol L^{-1}) affect the hydrocarbon oxidation with dioxygen catalyzed by transition metal complexes. The role of water as a ligand in the oxidation reactions catalyzed by metal complexes is also little studied (see Refs. [18], [23], and [24]). Only few examples of the effect of small amounts of water on the hydrocarbon oxidation catalyzed by metal compounds in the presence of onium salts are known [25, 26]. The oxidation depended on the concentration of ion pairs. Apparently, this was due to the changes in the structure of inverse micelles [26].

The effect of small concentrations of H_2O on the homolysis of hydroperoxides catalyzed by onium salts (including quaternary ammonium salts) was studied in more detail. For example, the catalytic effect of H_2O on the hydrocarbon oxidation in the presence of onium salts (QX) was considered [26]. Presumably, water molecules accelerated the ROOH decomposition due to the formation of H bonds involving H_2O, QX and ROOH molecules. This conclusion was confirmed by quantum chemical calculations [26].

Studying the effect of small amounts of water on the catalytic oxidation of hydrocarbons is also important in view of the fact that these processes always produce certain amount of water. It should be noted that the addition of low concentrations of water ($\sim 10^{-3}$ mol L^{-1}) to hydrocarbon solvents did not disturb the homogeneity of the medium [26].

8.3   THE EFFECT OF SMALL WATER CONCENTRATIONS ON ETHYLBENZENE OXIDATION WITH DIOXYGEN CATALYZED BY THE {FE^{III}(ACAC)$_3$ + R$_4$NBr} SYSTEM

The role of H-bonds in the mechanism of formation of catalytically active complexes was studied in our works by the example of the iron complex-mediated ethyl benzene oxidation with molecular oxygen in the presence of small amounts of water [27, 28].

It was assumed that this should have a positive effect on the conversion rate of iron complexes {Fe(acac)$_3$ + L^2} (L^2 = R$_4$NBr, 18C6) and also, probably, on the parameters S_{PEH} and C in this reaction.

It is well known that the stability of zwitterions increases in the presence of polar solvents. So outer sphere coordination of H$_2$O molecules may promote the stabilization of intermediate zwitterion L^2[L^1 M(L^1)$^{+}$O$_2$$^-$] and as a consequence the increase in the probability of the region selective addition of O$_2$ to nucleophilic g-C atom of (acac)$^-$ *ligand one expected.* The formation of hydrogen bonds of an H$_2$O molecule with a zwitterion can also favor the intra molecular proton transfer in the zwitterion, followed by conversion of the latter to the final products (see Schemes 8.1 and 8.2) [1, 2].

Also it is known, that insignificant amounts of water (about several millimoles) increased the yield of γ-C-alkylation products of the acac ion in an R$_4$N (acac) salt in reactions carried out in an aprotic solvent [29]. The catalytic activity of crown ethers (particularly, 18C6) in the reactions of electrophilic addition to the γ-C atom of an acetyl acetonate ligand in complexes M(acac)n carried out in THF was observed to increase upon addition of small amounts of water [29].

The positive effect of the introduction of small amounts of water (~ 10^{-3} mol L^{-1}) into the catalytic system {Fe^{III}(acac)$_3$+CTAB} on the selectivity of ethyl benzene oxidation to PEH was revealed for the first time in our studies [27].

It was found that the addition of small amounts of water leads to a nonadditive (synergistic) increase in the selectivity (SPEH)$_{max}$, the degree of conversion C and the parameter $\tilde{S} \cdot C$ as compared with their values in reactions of ethylbenzene oxidation catalyzed by Fe(acac)$_3$ and {Fe^{III}(acac)$_3$ + CTAB} in the absence of water (Figs. 8.1, 8.2). We used R$_4$NBr concentration in ~ the 10 times less than Fe(acac)$_3$ concentration to reduce the probability of formation of inverted micelles (in the case of CTAB). Onium salts QX are known to form with metal compounds complexes of variable composition depending on the nature of the solvent [2].

To assess the catalytic activity of nickel and iron systems {ML^1$_n$ + L^2} in ethyl benzene oxidation to a-phenyl ethyl hydro peroxide, a parameter $\tilde{S} \cdot C$ was chosen. $\tilde{S}$ is the averaged selectivity of ethyl benzene oxidation to PEH, which changes during the oxidation from S_0 (in the beginning of reaction) to a certain value S_{lim} (S_{lim} is an arbitrary value chosen as a standard); C is the degree of hydrocarbon conversion at $S = S_{lim}$ [1, 2]. For complexes of iron as S_{lim} a value $S_{lim}=(S_{PEH})_{max}=40\%$ in the reaction of ethyl benzene oxidation, catalyzed with $Fe^{II, III}(acac)_n$ at the absence of activated additives, was chosen (see kinetic curve 1 on Fig. 8.1).

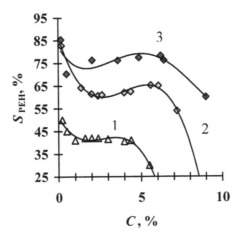

FIGURE 8.1 Dependences S_{PEH} vs. C in the ethyl benzene oxidation in the presence Fe(III)(acac)$_3$ (1) and systems {Fe(III)(acac)$_3$ + CTAB} (2) and {Fe(III)(acac)$_3$ + CTAB + H$_2$O} (3). [Fe^{III}(acac)$_3$] = 5×10^{-3} mol L^{-1}, [CTAB] = 0.5×10^{-3} mol L^{-1}, [H$_2$O]= 3.7×10^{-3} mol L^{-1}, 80 °C.

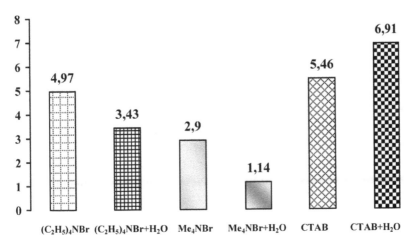

FIGURE 8.2 Parameter $\tilde{S} \cdot C \cdot 10^{-2}(\%,\%)$ in the ethyl benzene oxidation at catalysis systems {Fe^{III} (acac)$_3$+R$_4$NBr} and {Fe^{III} (acac)$_3$+R$_4$NBr+H$_2$O}. [Fe^{III}(acac)$_3$] = 5×10^{-3} mol L^{-1}, [R$_4$NBr] = 0.5×10^{-3} mol L^{-1}, [H$_2$O]= 3.7×10^{-3} mol L^{-1}, 80 °C.

The effect of water in small concentrations on the parameters of {Fe^{III} (acac)$_3$ + R_4NBr}-mediated ethyl benzene oxidation substantially depends on the structure of radical R in the cation of R_4NBr. Thus the introduction of ~10^{-3} mol L^{-1} of water to catalytic systems {$Fe(acac)_3+Me_4NBr$} and {$Fe(acac)_3+Et_4NBr$} led to a decrease in parameters $(SPEH)_{max}$ and $\tilde{S}\cdot C$. In this case (as for the catalysis with the {$Fe^{III}(acac)_{3+CTAB+H2}O$} system), the S_{PEH} vs. C dependence had a maximum.

The decrease in parameters $(S_{PEH})_{max}$ and $\tilde{S}\cdot C$ was explained by the high conversion rates of intermediates $Fe_x^{II}(acac)_y(OAc)_z(R_4NBr)_m(H_2O)_n$ to the final oxygenation products of $(Fe^{II}(acac)_2)_x(R_4NBr)_m(H_2O)_n$ complexes (R = Me, Et).

The increase in parameters $(S_{PEH})_{max}$, C and $\tilde{S}\cdot C$ observed in the ethyl benzene oxidation catalyzed by the {$Fe^{III}(acac)_3+CTAB+H_2O$} system was apparently associated with the increased steady-state concentration of selective heteroligand intermediates $Fe_x^{II}(acac)_y(OAc)_z(CTAB)_n(H_2O)_m$. The outer-sphere coordination of CTAB can give rise to steric hindrance for the coordination of H_2O molecules capable of accelerating the active complex decomposition. Moreover, a part of H_2O molecules can bind to the hydrophilic cation $Me_3(n-C_{16}H_{33})N^+$, which also leads to a decrease in the rate of conversion of active intermediates such as $Fe_x^{II}(acac)_y(OAc)_z(CTAB)_n(H_2O)_m$ into the final oxygenation products (see Scheme 8.2).

In the reaction of ethyl benzene oxidation in the presence of a {Fe (acac)$_3+(5\times10^{-4}$ mol $L^{-1})$ CTAB$+(3.7\times10^{-3}$ mol $L^{-1})H_2O$} system, no phenol was produced in the first 50 h after the reaction started. This can be explained by substantial reduction of the activity of catalytic complexes formed with respect to the PEH heterolysis and also by deceleration of formation of the complex Fe (OAc)$_2$ capable of accelerating the PEH heterolysis to PhOH that is inhibitor [27].

The ethyl benzene oxidation reactions under study follow the autocatalytic mechanism at 80 °C. After the auto-acceleration period associated with the transition of Fe^{III} ion to Fe^{II}, the reaction passed to the steady-state mode where $w = w_{max} = w_{lim}$ (w_0). The main oxidation products, namely, PEH, AP and MPC, were formed with an auto-acceleration period that increased with the addition of water. Under these conditions, the changes in the ethyl benzene oxidation rate were associated with the changes in the accumulation rates of PEH and/or AP and MPC [27]. The increase in w_0 observed in the catalysis by complexes $(Fe(acac)_2)_n(Me_4NBr)_m$ in the presence of 3.7×10^{-3} mol L^{-1} H_2O as compared with the catalysis by complexes $Fe(acac)_2$ and $(Fe(acac)_2)_n(Me_4NBr)_m$ in the absence of H_2O and only slight deceleration of ethyl benzene oxidation catalyzed by complexes $(Fe(acac)_2)_n(Et_4NBr)_m$ in the presence of H_2O as compared with its catalysis by complexes $(Fe(acac)_2)_n(Et_4NBr)_m$ are extraordinary facts [27, 30] when compared with the well-known observations of a considerable decrease in the hydrocarbon oxidation rate in the presence of water due to the solvatation of reactive radicals $RO_2^\times$ by water molecules [31] and also due to the catalyst deactivation with water formed in the course of the radical-chain oxidation of hydrocarbons with di oxygen in a nonpolar medium [32].

When ethyl benzene was catalyzed by the $\{Fe^{III}\ (acac)_3 + CTAB + H_2O\}$ system, a substantial decrease in the reaction rate w_0 (2-fold, Table 8.1) was observed.

It was found that the addition of water $(3.7 \times 10^{-3}$ mol $L^{-1})$ 'o catalytic systems $\{Fe^{III}\ (acac)_3 + R_4NBr\}$ $[R_4NBr = Me_3\ (n\text{-}C_{16}H_{33})\ NBr, Me_4NBr, Et_4NBr]$ did not change the mechanism of formation of the ethyl benzene main oxidation products. As in the absence of H_2O, throughout the reaction, three main products, namely, PEH, AP and MPC were formed in parallel $(w_{AP\,(MPC)}/w_{PEH} \neq 0$ at $t \to 0$, $w_{AP}/w_{MPC} \neq 0$ at $t \to 0)$. These results differ from the known facts of acceleration at of PEH decomposition when ethyl benzene oxidation was catalyzed by CTAB or catalytic systems based on CTAB and transition metal complexes as a result of PEH incorporation into the CTAB micelles [33, 34].

TABLE 8.1 The Rates (mol $L^{-1} \cdot s^{-1}$) of PEH and P Accumulations (P = $\{AP+MPC\}$) in the Presence of Systems $\{Fe^{III}\ (acac)_3 + R_4NBr\ (L^2)\}$ without Admixed $H_2O\ (L^3)$ and at the Addition of Small Amounts of $H_2O\ (3.7 \times 10^{-3}$ mol $L^{-1})$. $w_0\ \{(w_{PEH})_0 + (w_P)_0\}$. The Initial Rate of the Products Accumulation $w\ \{w_{PEH} + w_P\}$. The Rate of the Products Accumulation in the Course of the Ethyl Benzene Oxidation $[Fe^{III}\ (acac)_3] = 5 \times 10^{-3}$ mol L^{-1}. $[R_4NBr$ $(Me_4NBr, (C_2H_5)_4NBr, CTAB)] = 0.5 \times 10^{-3}$ mol L^{-1}–80 °C

L^2, L^3	$(w_{PEH})_0 \times 10^6$	$(w_P)_0 \times 10^6$	$w_{PEH} \times 10^6$	$w_P \times 10^6$
–	–	–	2.90 $(w_{PEH})_0 = w_{PEH}$	3.40 $(w_P)_0 = w_P$
$(C_2H_5)_4NBr$	5.9	8.3	3.3	4.3
$(C_2H_5)_4NBr$ $+ H_2O$	4.9	7.7	3.03	4.1
Me_4NBr	4.2	3.4	2.5	1.87
Me_4NBr $+ H_2O$	5.0	5.87	2.5	4.8

TABLE 8.1   *(Continued)*

L^2, L^3	$(w_{PEH})_0 \times 10^6$	$(w_p)_0 \times 10^6$	$w_{PEH} \times 10^6$	$w_p \times 10^6$
CTAB	4.35	3.3	2.5	1.2
CTAB + H$_2$O	3.75	1.1	3.19	0.85

8.4   THE EFFECT OF SMALL CONCENTRATIONS OF WATER ON ETHYLBENZENE OXIDATION WITH DIOXYGEN CATALYZED BY THE {FE^{III} (ACAC)$_3$ + 18C6} SYSTEM

In the reaction of ethyl benzene oxidation with dioxygen in the presence of a catalytic system {Fe(acac)$_3$+18C6} {[Fe(acac)$_3$] = 5×10^{-3} mol L^{-1}, [18C6] = 5×10^{-3} or 5×10^{-4} mol L^{-1}} at 80 °C, the S_{PEH} vs. C dependence had a maximum (Fig. 8.3). For the same degrees of ethyl benzene conversion in the oxidation reactions, the values $(S_{PEH})_{max}$ corresponding to the catalysis by the {Fe(acac)$_3$ + 18C6} system (70% for [18C6]$_0$=5×10^{-4} mol L^{-1} and 75.7% for [18C6]$_0$= 5×10^{-3} mol L^{-1}) exceeded $(S_{PEH})_{max}$ (65%) observed in the catalysis by CTAB and far exceeded S_{PEH} (40%) corresponding to the catalysis by the complex Fe(acac)$_3$ in the absence of 18C6 [$w = w_{max} = w_{lim}(w_0)$]. With the increase in S_{PEH} from 40% [catalysis by the complex Fe (acac)$_3$] to 70%–75.7% [catalysis by the {Fe(acac)$_3$+18C6} system], the conversion increased twofold, namely, from 4% to 8% (or 10%) [35].

The addition of 18C6 to the catalyst Fe(acac)$_3$ changed the composition of the oxidation products. A substantial increase in Ref. [PEH]$_{max}$ (~1.6- or 1.7-fold for [18C6] = 5×10^{-4} or 5×10^{-3} mol L^{-1}, respectively) and a considerable decrease in the AP and MPC concentrations (4.5-fold) were observed. Moreover, the addition of 18C6 resulted in substantial inhibition of the PEH heterolysis to phenol and also in the increase in parameter Š·C (2.5- and 2.8-fold for [18C6] = 5×10^{-4} and 5×10^{-3} mol L^{-1}, respectively) as compared with the value Š·C observed in the catalysis by the complex Fe (acac)$_3$ (Fig. 8.4) [35].

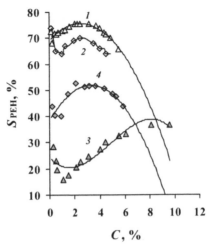

FIGURE 8.3 Dependence of S_{PEH} vs. C in the reactions of the oxidation of ethyl benzene catalyzed catalytic systems {Fe^{III} (acac)$_3$+18C6 (5×10^{-3} mol L^{-1})} and {Fe^{III} (acac)$_3$+18C6(0.5×10^{-3} mol L^{-1})} without additives of water (1, 2), and in the presence of 3.7×10^{-3} mol L^{-1} H$_2$O. [Fe^{III} (acac)$_3$]=5×10^{-3} mol L$^{-1-80\,°C}$.

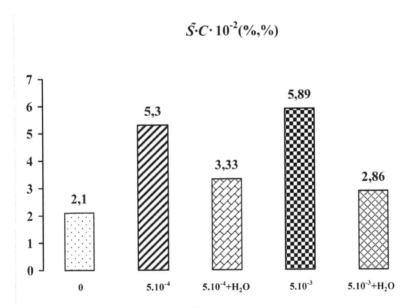

FIGURE 8.4 The values of parameter $\tilde{S}\cdot C \cdot 10^{-2}$ (%, %) in the reactions of the oxidation of ethyl benzene catalyzed by Fe^{III} (acac)$_3$ ([18C6] =0) or catalytic systems {Fe^{III} (acac)$_3$+18C6 (0.5×10^{-3} or 5×10^{-3}(mol L^{-1}))} without additives of water, and in the presence of 3.7×10^{-3} mol L^{-1} H$_2$O. [Fe^{III} (acac)$_3$]=5×10^{-3} mol L^{-1}–80 °C.

The synergistic effect of the increase in parameters S_{PEH} and $\tilde{S}\cdot C$, observed in ethyl benzene oxidation catalyzed by {Fe(acac)$_3$ + 18C6} systems [36], suggests that catalytically active complexes (Fe(acac)$_2$)$_p$(18C6)$_q$ and their transformation products were formed during the oxidation. For example, Fe^{II} and Fe^{III} halides are known to form complexes with crown ethers with different compositions (1: 1, 1: 2, 2: 1) and structures, depending on the nature of the crown ether and solvent [36]. Due to the favorable combination of electronic and steric factors that operate at the inner-and outer-sphere coordination (hydrogen bonds) of 18C6 with Fe(acac)$_2$ stable active poly nuclear hetero ligand complexes with the composition Fex^{II}(acac)$_y$(OAc)$_z$(18C6)$_n$ are formed with the high probability, which in turn favors the increase in parameters $(S_{PEH})_{max}$ and C.

$$\{Fe\,(acac)_3 + 18C6\} \rightarrow [Fe^{II}(acac)_2]_{px}(18C6)_q \xrightarrow{\text{o2}} Fe_x^{II}(acac)_y(OAc)_z(18C6)_n.$$

With the addition of water (3.7×10^{-3} mol L^{-1}), the efficiency of catalytic systems {Fe(acac)$_{3+18C6}$} specified by parameters $(SPEH)_{max}$ and $\tilde{S}\cdot C$ decreased; however, the conversion increased from 4% to ~6.5% (for [18C6] = 5×10^{-4} mol L^{-1}) or 9%(for[18C6] = 5×10^{-3} mol L^{-1}) (compare Figs. 8.9, 8.11 and 8.12) [27, 28].

An analogous decrease in parameters $(S_{PEH})_{max}$ and $\tilde{S}\cdot C$ was also observed for the ethyl benzene oxidation with dioxygen catalyzed by {Fe(acac)$_3$ + R$_4$NBr} (R = Me, Et) systems in the presence of small portions of water (Fig. 8.2) [27].

Coordination of water molecules with complexes [Fe^{III,II}(acac)$_n$]$_p$×(18C6)$_q$ to form complexes [Fe^{II}(acac)$_2$]$_x$(18C6)$_y$(H$_2$O)$_m$ and the transformation of the latter during the ethyl benzene oxidation into active Fe$_x^{II}$(acac)$_y$(OAc)$_z$(18C6)$_n$(H$_2$O)$_m$ species may explain the observed trends in variations of parameters $S_{PEH,}$ C and $\tilde{S}\cdot C$ in the presence of small H$_2$O concentrations (see Figs. 8.10–8.13). Apparently, coordination of water molecules to Fe complex does not lead to 18C6 displacement to the second coordination sphere [28, 37], because no decrease in the activity of the {Fe (acac)$_3$+18C6} system with respect to oxidation was observed. It is also well-known that water molecules can form molecular complexes with crown ethers through the formation of hydrogen bonds (insertion complexes) [38], however, the enthalpy of their formation is low (~2–3 kcal mol^{-1}) [39].

Apparently, the outer-sphere coordination of water molecules with the complex [Fe^{II}(acac)$_2$]$_x$×(18C6)$_q$ favors the conversion of the latter into active species such as Fe$_x^{II}$(acac)$_y$(OÁc)$_z$(18C6)$_n$(H$_2$O)$_m$ capable of catalyzing ethyl benzene oxidation to PEH, as evidenced by the increase in the initial rates of ethyl benzene oxidation (see Table 8.2) and the increase in S_{PEH} in the course of reaction (see Fig. 8.2, compare with the facts of the increase in the oxidation rates in the case of the catalysis by systems $\{Fe^{III}(acac)_3+Me_4NBr+H_2O\}$.

The decrease in parameter $(S_{PEH})_{max}$ can apparently be explained by the decrease in the steady-state concentration of reactive intermediates $Fe_x^{II}(acac)_y(OAc)_z(18C6)_n(H_2O)_m$ in the course of oxidation and the acceleration of their transformation into the final product, catalyst $Fe(OAc)_2$.

It was established important fact. In the presence of water in small concentrations, the ratio of main products of ethyl benzene oxidation changed. In the beginning of this reaction ($C \leq 1.0\%$) catalyzed by the $\{Fe(acac)_3+18C_6+H_2O\}$ system, acetophenone rather than PEH was the main product ([AP] > [PEH] and [AP]:" [MPC]). The selectivity of ethyl benzene oxidation to acetophenone $[(S_{AP})_0]$ increased from 24 (without added H_2O) to ~70% (with added H_2O), where $(S_{PEH})_0 \approx 25\%$, and the [AP]: [MPC] ratio increased from 1.2 (in the absence of added H_2O) to 6.5 (in the presence of H_2O) (Figs. 8.5 and 8.6).

With regard to the mechanism for the formation of main products, it was discovered that as in catalysis with system $\{Fe(acac)_3+CTAB\}$, introduction of small portions of water $(3.7\times10^{-3}$ mol $L^{-1})$ to the $\{Fe(acac)_3+18C6\}$ systems did not affect the mechanism of formation of the main oxidation products. Namely, throughout the ethyl benzene oxidation, three main products, PEH, P (P=AP or MPC) were formed in parallel $_(w_P/w_{PEH}$ ^{1} 0 and w_{AP}/w_{MPC} ^{1} 0 at t → 0).

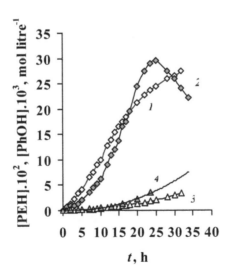

FIGURE 8.5   Kinetics of accumulation of PEH (1, 2) and PhOH (3, 4) in the reactions of the oxidation of ethyl benzene catalyzed $\{Fe(acac)_3+18C6\}$ (1,3) and $\{Fe(acac)_3+18C6+H_2O\}$ (2, 4) $[Fe(acac)_3]=[18C6]=5\times10^{-3}$ mol L^{-1}. $[H_2O]=3.7\times10^{-3}$ mol L^{-1} –80 °C.

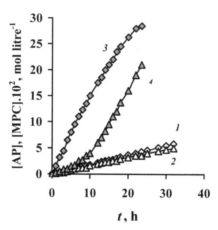

FIGURE 8.6   Kinetics of accumulation of AP (1, 3) и MPC (2, 4) in the reactions of the oxidation of ethyl benzene catalyzed by systems {Fe (acac)$_3$+18C6} (1, 2) and {Fe(acac)$_3$+18C6+H$_2$O} (3, 4). [Fe (acac)$_3$]= [18K6] =5×10^{-3} mol L^{-1} [H$_2$O] =3.7×10^{-3} mol L^{-1}–80 °C.

So, the introduction of small portions of water (~10^{-3} mol L^{-1}) to catalytic systems based on iron complexes (5.0×10^{-3} mol L^{-1}) leads to the higher catalytic activity of these systems: the main thing, the growth of $S_{PEH,max}$ (from 65 to 78%) (L^2=CTAB (5.0×10^{-4} mol L^{-1})), and besides this the increase in the initial rate (L^2=Me$_4$NBr (5.0×10^{-4} mol L^{-1})), the growth of the initial rate and high selectivity (C=1%) on acetophenone S_{AP} =70% (L^2=18C6).

8.5   ASSESSING THE ACTIVITY OF FE-CATALYSTS IN MICRO STEPS OF RADICAL-CHAIN ETHYL BENZENE OXIDATION WITH DIOXYGEN ROLE OF H-BONDING

An original method was developed for assessing the catalytic activity of metal complexes in the micro steps of chain initiation (Eq. (1)) and propagation (Eq. (2)) with the use of a simplified scheme. Taking into account that complexes ML^1 {M = Ni^{II} ([Cat] = (0.5–1.5) ×10^{-4} mol L^{-1}) and Fe^{II,III} ([Cat] = (0.5–5) ×10^{-3} mol L^{-1})} are inactive in the PEH homolysis [1, 2], the rate of homolytic decomposition of hydro peroxide can be equated to zero [1, 2, 40–42]. With the assumption relationships w_0~[Cat]$^{1/2}$ and $(w_i)_0$~[Cat] established for the radical chain ethyl benzene oxidation catalyzed by complexes ML$^1_2$ are also fulfilled for the catalysis by complexes ML$^1_2$× L^{2}containing activating ligands, then it is possible to ignore the linear radical termination in the catalyst and take into account only quadratic termination [1, 2].

In this case, the rate of quadratic chain termination can be calculated using equation

$$\text{wterm} = k5[RO2\times]2 = k5\left\{\frac{w_{PEH}}{k_2[RH]}\right\}^2 \tag{1}$$

where w_{PEH} is the rate of PEH accumulation, k_5 is the rate constant for quadratic chain termination, k_2 is the rate constant for chain propagation $RO_2\times$ $+RH\rightarrow$.

The w_{term} values determined using Eq. (1) for quasi-steady state conditions with respect to RO_2 radicals and designated as w_i are the measure of the metal activity with respect to molecular O_2 The deviation between values $wAP+MPC$ and w_{term} in the absence of linear termination of chains on the catalyst was attributed to the additional formation of an alcohol and a ketone in the step of chain propagation at participation of Cat ($Cat+RO_2\times\rightarrow$). The rate of the chain propagation step

$$w_{pr} = w_{AP+MPC} - w_{term} \tag{2}$$

The involvement of nickel and iron complexes in the chain propagation step is testified by the direct proportional dependence w_{pr} vs. [Cat].

The involvement of transition metal complexes in the chain propagation step of ethyl benzene oxidation was considered exclusively in our studies [1, 2, 40–42] and in theoretical publications [43, 44].

The schemes of radical chain oxidation of hydrocarbons that involve the transient formation of peroxide complexes [LM–OOR] [45–47] followed by their homolytic decomposition ([LM–OOR]→R¢C=O (ROH) + R^x) allow one to explain the parallel formation of an alcohol and a ketone during the ethyl benzene oxidation in the presence of complexes ML_n^1 and ML_n^1 with L^2 [M = Fe^{II,III}; L^1 = acac; n = 2, 3; L^2 = DMF, HMPA, R$_4$NBr, 18C6].

8.5.1 MECHANISM OF CATALYSIS WITH IRON COMPLEXES

Chain initiation in the ethyl benzene oxidation with molecular oxygen in the presence of complexes Fe (acac)$_3$ or {Fe(acac)$_3$+L^2} can be described by the following scheme:

$$\text{Fe}^{III}\text{ (acac)}_3 + RH \rightarrow \text{Fe}^{II}\text{ (acac)}_2\text{xxxHacac} + R^x \text{ or}$$

$$(\text{Fe}^{III}(\text{acac})_3)_m \cdot L^2_n + RH \rightarrow (\text{Fe}^{II}(\text{acac})_2)_x(L^2)_y\text{xxxHacac} + R^{x.}$$

The subsequent reaction of the FeII complexes formed with dioxygen leads to chain initiation $[(w_i)_0]$ in a region where $w = w_{max} = w_{lim}$ (w_0), that is, after the end of the induction period associated with the Fe^{III} conversion to Fe^{II} and to the transition into the steady-state reaction mode.

The rate of chain initiation in the ethyl benzene oxidation in the presence of complexes {Fe (acac)$_3$+L^2} (L^2 = R$_4$NBr, HMPA, DMF) is higher than in the presence of Fe (acac) 3 (see, e.g., Table 8.2) and much higher than in the noncatalytic oxidation of ethyl benzene $[(w_i)_0 \approx 10^{-9}$ mol L^{-1} s$^{-1}]$.

For the ethyl benzene oxidation catalyzed by complexes {Fe (acac)$_3$ + CTAB}, $(S_{PEH})_0$ depends on the catalyst activity in the chain initiation and propagation steps. Thus, the increase in $(S_{PEH})_0$ from 45% to ~65% was associated with the twofold increase in $(w_i)_0$ [this was accompanied by an insignificant decrease in $(w_{pr})_0$] (see Table 8.2). The rate of PEH accumulation in the presence of {Fe (acac)$_3$ + CTAB} compared to the catalysis with Fe (acac)$_3$ increased 1.5 fold and the rate of AP and MPC formation decreased. Similar results were also obtained in the presence of {Fe (acac)$_3$+Me$_4$NBr}. However, when {Fe (acac)3+Et$_4$NBr} served as the catalyst of ethyl benzene oxidation, $(S_{PEH})_0$ was observed to decrease because the increase in $(w_{AP}$+MPC)$_0$ much exceeded the gain in $(w_{PEH})_0$.

The increase in $(S_{PEH})_0$ from 40%–50% to 65%–70% in the initial stages of ethyl benzene oxidation catalyzed by complexes $(Fe^{II}(acac)_2)_p(18C6)_g$ was largely associated with the deceleration of AP and MPC formation in the micro step of chain propagation. Apparently, due to the steric factors and H-bonds that arise at the inner or outer-sphere coordination of 18C6 with Fe^{II} (acac)$_2$ the probabilities of coordination of oxygen $[(w_i)_0]$ and/or radicals RO$_2^x$ $[(w_{pr})_0]$ with the metal ions to form the primary catalyst complexes with an O$_2$ molecule and/or a radical RO$_2^x$, followed by the formation of active species of the superoxide or peroxide types, decreased.

The ratio $[(w_i)_0/(w_{pr})_0]$ 100% $\approx$ 2%–5% means that iron complexes {Fe(acac)$_3$+L^2} are more active in the chain propagation step than in the step of chain initiation. The products AP and MPC were largely formed in the chain propagation step probably due to the homolytic decomposition of the transient complex $[L^2FeL^1_2-OOR]$ [48–51].

Complexes $Fe^{II}(acac)_2 \times$ HMPA formed in the initial stages of ethyl benzene oxidation in the presence of the {$Fe^{III}(acac)_3$ + HMPA} catalytic system remain unchanged during the reaction [2]. The dependences of the reaction rate w_0, of calculated values $(w_{pr})_0$ and $(w_{pr})_0$, and also of the parameter $\tilde{S}\cdot C$ on the HMPA concentration for the ethyl benzene oxidation with dioxygen (80 °C) under steady-state conditions had extreme. The maximum values of these parameters were reached for the same HMPA concentration $[\sim(2.5–5.0) \times 10^{-2}$ mol L$^{-1}]$. Apparently, this concentration corresponds to the formation of active complexes $Fe^{II}(acac)_2 \times HMPA$ with the 1:1 composition and argues for the proposed method of assessing the values $(w_{pr})_0$ and $(w_{pr})_0$.

Coordination of HMPA with $Fe^{II}(acac)_2$ results in a decrease in the effective activation energies (E_a) of micro steps of chain initiation and propagation by 48.15 and 15.9kJ mol^{-1}, respectively, as compared with E_a values obtained in the catalysis by complexes $Fe^{II}(acac)_2$ in the absence of HMPA. The gain in the activation energy of chain initiation observed at the HMPA coordination was equal to the energy consumed in the addition of this extraligand to metal acetyl acetonates (~41.87 kJ mol^{-1}) [48]. As the temperature decreased from 120 to 80 °C, the difference in E_a between the reactions of chain initiation (54.55 kJ mol^{-1}) and propagation (73.81 kJ mol^{-1}) in the presence of HMPA can, apparently, explain the increase in the selectivity of ethyl benzene oxidation to PEH from 46% (at 120 °C) to 57% (at 80 °C).

The higher catalytic activity of complexes $Fe(acac)_2 \times DMF$ as compared with $Fe(acac)_2 \times HMPA$ in the steps of chain initiation and propagation in ethyl benzene oxidation is apparently associated with the π-donating properties of DMF and its trend to H-bonding [49]. It is quite probable that the coordination of a π-donating ligand (DMF) with $Fe(acac)_2$ favors the formation of primary complexes of $Fe(acac)_2 \times DMF$ with oxygen [50, 51] of the superoxide type, namely, $[DMF \times Fe(acac)_2 \times O_2^-]$, and also the increase in their activity in the step of radical formation. Schemes of radical chain oxidation of hydrocarbons that include the transient formation of peroxo complexes [LM-OOR] in the chain propagation step attribute the increase in ethyl benzene oxidation rate and selectivity with respect to MPC $[(S_{MPC})_0 \approx 58\%]$ in the presence of a catalytic complex $Fe(acac)_2 \times DMF$. The additional coordination of Fe with a DMF ligand favors the stabilization of oxo-species $[DMF \times Fe^{II}-O^\times]$ formed in the homolytic decomposition of peroxo complexes $[DMF \times Fe-OOR]$ at the O−O bond $[L^2Fe^{II}-O \times - \times OR] \rightarrow [L^2Fe^{II}-O \times] + [\times OR] \rightarrow R'R''C=O$ (or ROH) + R, and the probability of escape of radicals RO× from the solvent cage (the inner-cage 'latent-radical' mechanism).

Above, it was also assumed that the mechanism of ethyl benzene oxidation catalyzed by complexes $\{Fe(acac)_3 + R_4NBr\}$ or $\{Fe(acac)3+18C6\}$ remains unchanged upon the introduction of small portions of water (3.7×10^{-3} mol L^{-1}). In this case too, the conditions $w_0 \sim [Cat]^{1/2}$ and $(w_i)_0 \sim [Cat]$ are fulfilled, which allows one to assess the activity of complexes $(Fe^{II}(acac)_2)_x(R_4NBr)_y(H_2O)_n$ and $(Fe^{II}(acac)_2)_x(18C6)_y(H_2O)_n$ in micro steps of chain initiation and propagation.

In the reaction of ethyl benzene oxidation catalyzed by complexes $(Fe^{II}(acac)_2)_x(CTAB)_y(H_2O)_n$ (see Table 8.2), the deceleration of reaction in initial oxidation stages was associated with a considerable (3.2-fold) decrease in the chain propagation rate $(w_{pr})_0$. On the other hand, $(w_i)_0$ decreased only ~1.3-fold; hence, the ratio $[(w_i)_0/(w_{pr})_0] \times 100\%$ increased by a factor of 2.35. This enhanced the selectivity $(S_{PEH})_0$ in the initial reaction stages. The decrease in the chain propagation rate $(w_{pr})_0$ can be explained by the steric hindrance posed by a CTAB ligand to the coordination of radicals $RO_2{}^\times$ with the central metal ion.

TABLE 8.2   The Initial Rates w_0 and Calculated Rates of Micro Stages of Chain Initiation $(w_i)_0$, Chain Propagation $(w_{pr})_0$ (in mol L^{-1} s^{-1}) and $[(w_i)_0/(w_{pr})_0] \times 100\%$ in the Ethyl Benzene Oxidations Catalyzed with $Fe^{III}(acac)_3$ or $\{Fe^{III}(acac)_3+L^2\}$, or $\{Fe^{III}(acac)_3+L^2+L^3\}$ systems. $[Fe^{III}(acac)_3]=5\times10^{-3}$ mol L^{-1}. $L^2=[CTAB]=0.5\times10^{-3}$ mol L^{-1}. $L^2=[18C6] = 5\times10^{-3}$ mol L^{-1}. $L^3= [H_2O] = 3.7\times10^{-3}$ mol L^{-1} –80°C.

L^2, L^3	$w_0\times10^6$	$(w_i)_0\times10^7$	$(w_{pr})_0\times10^6$	$[(w_i)_0/(w_{pr})_0] \times100\%$
–	6.30	0.79	3.32	2.38
CTAB	7.65	1.63	3.14	5.19
CTAB+H_2O	4.85	1.21	0.98	12.24
18C6	2.63	0.24	0.93	2.58
18C6 + H_2O	6.94	0.15	5.58	0.27

In contrast to the ethyl benzene oxidation catalyzed by complexes $(Fe^{II}(acac)_2)_x(CTAB)_y(H_2O)_n$, for its catalysis by complexes $(Fe^{II}(acac)_2)_x(R_4NBr)_y(H_2O)_n$ (R = Me, Et), the $[(w_i)_0/(w_{pr})_0] \times 100\%$ ratio decreased by a factor of ~ 1.4 (R = Et) or ~ 1.22 (R = Me) as compared with this ratio calculated for the reaction catalyzed by complexes $(Fe^{II}(acac)_2)_x(R_4NBr)_y$, without addition of H_2O.

As follows from the data shown in Table 8.2, the reaction of chain propagation is the main source of AP and MPC at the ethyl benzene oxidation catalyzed by complexes $(Fe^{II}(acac)_2)_x(L^2)_y(H_2O)_n$ $(L^2 = R_4NBr, 18C6)$. The contribution of quadratic chain termination is insignificant. An analogous result was obtained for the catalysis of ethyl benzene oxidation by complexes $(F_e{}^{II}(acac)_2)_m(R_4NBr)_n$, $(Fe^{II}(acac)_2)p(18C6)_q$ or $Fe^{II,III}(acac)_n$ [2].

Introduction of water into the reaction of ethyl benzene oxidation catalyzed by complexes $\{Fe(acac)_3+18C6\}$ causes a substantial (10-fold) decrease in the $[(w_i)_0/(w_{pr})_0] \times100\%$ ratio largely due to the increase in $(w_{pr})_0$. When the $\{Fe(acac)_3+18C6+H_2O\}$ system was used as the catalyst, the lowest selectivity of ethyl benzene oxidation to PEH was observed in the initial stages $[(S_{PEH})_0 \approx 25\%]$, where the highest selectivity corresponded to the ethyl benzene oxidation to acetophenone $[(S_{AP})_0 = 70\%]$. The [AP]/[MPC] ratio increased from ~ 1.2 (for the catalysis with the $\{Fe(acac)_3 + 18C6\}$ system) to 6.5 (for the catalysis with the $\{Fe(acac)_3 + 18C6 + H_2O\}$ system).

So, studying the effect of addition of small amounts of water to catalytic systems {FeL^{1}n + L^2} confirmed the important role of H-bonds in the formation of catalytic species. It was found that the introduction of small portions of water to catalytic systems based on iron complexes leads to the higher catalytic activity of these systems.

The ethyl benzene oxidation (70°C, CH_2Cl_2, CH_3CHO) upon catalysis by complexes of Cu(II) with 18C6 [$(CuCl_2)_4(18C6)_2(H_2O)$] afforded acetophenone as the main product with the formation of ketone (AP). The [ketone]/[alcohol] ration is 6.5 [52]. Similar results were obtained for oxidation of other hydrocarbons, for example, indane, cyclohexane, and tetralin.

The selectivity of radical-chain oxidation of ethyl benzene to acetophenone in the presence of {Fe(acac)$_3$ + 18C6 + H_2O} was sufficiently high but lower than in the catalysis by systems mimicking the effect of mono oxygenases, for example, the Sawyer system, that is, {[Fe^{II}(Mn^{III})] + L$_x$ + HOOH(HOOR) + Solv} [L$_x$ = bipy, py; Solv are polar solvents (MeCN, py + AcOH)] [53, 54]. In the ethyl benzene oxidation with dioxygen (24 °C) catalyzed by the Sawyer system, the selectivity with respect to acetophenone reached 100% but for very low hydrocarbon conversion of 0.4%.

8.6 ROLE OF SUPRAMOLECULAR NANOSTRUCTURES FORMATION DUE TO H-BONDING IN MECHANISM OF CATALYSIS WITH CATALYTIC ACTIVE COMPLEXES FE$_x$(ACAC)$_{n}^2$ $_M$(H$_2$O)$_N$ (L^2 = 18C6), FE$_x$(ACAC)$_{n}^2$ $_M$ (L^2 = CTAB)

Hydrogen bonds vary enormously in bond energy from ~15–40 kcal/mol for the strongest interactions to less than 4 kcal/mol for the weakest. It is proposed, largely based on calculations, that strong hydrogen bonds have more covalent character, whereas electrostatics are more important for weak hydrogen bonds, but the precise contribution of electrostatics to hydrogen bonding is widely debated [55]. Hydrogen bonds are important in noncovalent aromatic interactions, where π-electrons play the role of the proton acceptor, which are a very common phenomenon in chemistry and biology. They play an important role in the structures of proteins and DNA, as well as in drug receptor binding and catalysis [56]. Proton-coupled bicarboxylates top the list as the earliest and still the best-studied systems suspected of forming low-barrier hydrogen bonds (LBHBs) in the vicinity of the active sites of enzymes [57].

$$R-\overset{\overset{\displaystyle O}{\|}}{C}-O^-\text{---}H-O-\overset{\overset{\displaystyle O}{\|}}{C}-R'$$

These hydrogen-bonded couples can be depicted as and they can be abbreviated by the general formula $X^-\times\times\times HX$. Proton-coupled bicarboxylates appear in 16% of all protein X-ray structures. There are at least five X-ray structures showing short (and therefore strong) hydrogen bonds between an enzyme carboxylate and a reaction intermediate or transition state analog bound at the enzyme active site. The authors [57] consider these structures to be the best de facto evidence of the existence of low-barrier hydrogen bonds stabilizing high-energy reaction intermediates at enzyme active sites. Carboxylates figure prominently in the LBHB enzymatic story in part because all negative charges on proteins are carboxylates.

Nanostructure science and supra molecular chemistry are fast evolving fields that are concerned with manipulation of materials that have important structural features of nanometer size (1 nm to 1 μm) [58]. Nature has been exploiting no covalent interactions for the construction of various cell components. For instance, microtubules, ribosomes, mitochondria, and chromosomes use mostly hydrogen bonding in conjunction with covalently formed peptide bonds to form specific structures.

H-bonds are commonly used for the fabrication of supra molecular assemblies because they are directional and have a wide range of interactions energies that are tunable by adjusting the number of H-bonds, their relative orientation, and their position in the overall structure. H-bonds in the center of protein helices can be 20 kcal/mol due to cooperative dipolar interactions [59, 60].

The porphyrin linkage through H-bonds is the binding type generally observed in nature. One of the simplest artificial self-assembling supramolecular porphyrin systems is the formation of a dimer based on carboxylic acid functionality [61].

As mentioned before, in the effect of Fe^{II}-acetylacetonate complexes, we have found the analogy with the action of Fe^{II}-Acireductone Dioxygenase (ARD) and with the action of Fe^{II} containing Acetyl acetone Dioxygenase (Dke1) [11,14].

For iron complexes oxygen adds to C–C bond (rather than inserts into the C=C bond as in the case of catalysis with nickel (II) complexes) to afford intermediate, that is, a Fe complex with a chelate ligand containing 1,2-dioxetane fragment (Scheme 8.2). The process is completed with the formation of the (OAc) ⁻ chelate ligand and methylglyoxal as the second decomposition product of a modified acac-ring (as it has been shown in Ref. [14]).

The methionine salvage pathway (MSP) (Scheme 8.3) plays a critical role in regulating a number of important metabolites in prokaryotes and eukaryotes. Acireductone Dioxygenases (ARDs) Ni(Fe)-ARD are enzymes involved in the methionine recycle pathway, which regulates aspects of the cell cycle. The relatively subtle differences between the two metalloproteins complexes are amplified by the surrounding protein structure, giving two enzymes of different structures and activities from a single polypeptide (Scheme 8.3) [62]. Both enzymes $Ni^{II}(Fe^{II})$-ARD are members of the structural super family, known as cupins, which also include Fe-Acetyl acetone Dioxygenase (Dke1) and Cysteine Dioxygenase. These enzymes that form structure super family of cupins use a triad of histidine-ligands (His), and also one or two oxygens from water and a carboxylate oxygen (Glu), for binding with Fe-center [63].

Structural and functional differences between the two ARDs enzymes are determined by the type of metal ion bound in the active site of the enzyme.

The two aci-reductone dioxygenase enzymes (ARD and ARD') share the same amino acid sequence, and only differ in the metal ions that they bind, which results in distinct catalytic activities. ARD has a bound Ni^{+2} atom while ARD' has a bound Fe^{+2} atom. The apo-protein, resulting from removal of the bound metal, is identical, and is catalytically inactive. ARD and ARD' can be interconverted by removing the bound metal and reconstituting the enzyme with the alternative metal. ARD and ARD' act on the same substrate, the aci-reductone, 1,2-dihydroxy-3-keto-5-methylthiopentene anion, but they yield different products. ARD' catalyzes a 1,2-oxygenolytic reaction, yielding formate and 2-keto-4-methylthiobutryate, a precursor of methionine, and thereby part of the methionine salvage pathway, while Ni-ARD catalyzes a 1,3-oxygenolytic reaction, yielding formate, carbon monoxide, and methylthioproprionate, an off-pathway transformation of the aci-reductone. The role of the ARD catalyzed reaction is unclear.

We assumed that one of the reasons for the different activity of Ni^{II} (Fe^{II})-ARD in the functioning of enzymes in relation to the common substrates (Acireductone (1,2-Dihydroxy-3-keto-5-methylthiopentene-2) and O_2) can be the association of catalyst in various macrostructure due to intermolecular H-bonds.

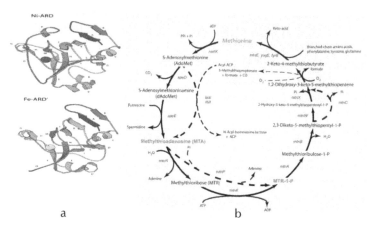

a   b

SCHEME 8.3   Acireductone Dioxygenases Ni-ARD and Fe-ARD' (a) [62] are involved in the methionine recycle pathway (b).

The Fe^{II}ARD operation seems to comprise the step of oxygen activation (Fe^{II}+O$_2$→ Fe^{III}–O$_2^-$×) (by analogy with Dke1 action [14]). Specific structural organization of iron complexes may facilitate the following regioselective addition of activated oxygen to Acireductone ligand and the reactions leading to formation of methionine. Association of the catalyst in macrostructures with the assistance of the intermolecular H-bonds may be one of reasons of reducing Ni^{II}ARD activity in mechanisms of Ni^{II}(Fe^{II})ARD operation [62, 63]. Here for the first time we demonstrate the specific structures organization of functional model iron complexes in aqueous and hydrocarbon medium.

The possibility of the formation of stable supramolecular nanostructures on the basis of iron (nickel) heteroligand complexes due to intermolecular H-bonds we have researched with the AFM method [64].

First we received UV-spectrum data, testified in the favor of the complex formation between Fe(acac)$_3$ and 18-crown-6, 18C6, that modeled ligand surrounding Fe-enzyme. Here for the spectrums of solutions of Fe (acac)$_3$ (1) and mixture {Fe (acac)$_3$+18C6} (2) in various solvents are presented.

As one can see in at the addition of the 18C6 solution (in CHCl$_3$) to the Fe (acac)$_3$ solution (in CHCl$_3$) (1:1) an increase in maximum of absorption band in spectrum for acetylacetonate-ion (acac)$^-$ in complex with iron, broadening of the spectrum and a bathochromic shift of the absorption maximum from λ ~ 285 nm to λ = 289 nm take place. The similar changes in the intensity of the absorption band and shift of the absorption band are characteristic for narrow, crown unseparated ion-pairs [2]. Earlier similar changes in the UV- absorption band of Co^{II}(acac)$_2$ solution we observed in the case of the coordination of macrocyclic polyether 18C6 with Co^{II}(acac)$_2$ [2]. The formation of a complex between Fe(acac)$_3$ and 18C6 occurs at preservation of acac ligand in internal coordination sphere of Fe^{III} ion because at the another case the short-wave shift of the absorption band should be accompanied by a significant increase in the absorption of the solution at $\lambda = 275$ nm, which correspond to the absorption maximum of acetyl acetone. It is known that Fe^{II} and Fe^{III} halogens form complexes with crown-ethers of variable composition (1:1, 1:2, 2:1) and structure dependent on type of crown-ether and solvent [36]. It is known that Fe(acac)$_3$ forms labile OSCs (Outer Sphere Complexes) with CHCl$_3$ due to H-bonds [23].

However in an aqueous medium the view of UV-spectrum is changing (Fig. 8.7b): an decrease in absorption maximum of acetylacetonate ion (acac)$^-$ (in Fe(acac)$_3$) at the addition of a solution of 18C6 to the Fe(acac)$_3$ solution (1:1). Possibly, in this case inner-sphere coordination of 18C6 cannot be excluded.

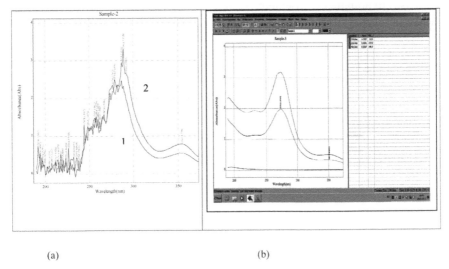

(a)   (b)

FIGURE 8.7   Absorption spectra of iron complexes: (a) Fe(acac)$_3$ (1, red), mixture {Fe(acac)$_3$ + 18C6}(1:1) in CHCl$_3$ (2, blue); (b) Fe(acac)$_3$ (red) and mixture {Fe(acac)$_3$ + 18C6}(1:1) (blue)- in H$_2$O, 20 °C.

In an aqueous medium the formation of supramolecular structures of generalized formula Fe$^{III}_x$(acac)$_y$18C6$_m$(H$_2$O)$_n$ is quite probable.

In the Figs. 8.8 and 8.9, three-dimensional and two-dimensional AFM image of the structures on the basis of iron complex with 18C6 $Fe^{III}_x(acac)$ $_y18C6_m(H_2O)_n$, formed at putting a uterine solution on a hydrophobic surface of modified silicone are presented. It is visible that the generated structures are organized in certain way forming structures resembling the shape of tubule micro fiber cavity (Fig. 8.9c). The heights of particles are about 3–4 nm. In control experiments it was shown that for similar complexes of nickel $Ni^{II}(acac)_2\times18C6\times(H_2O)_n$ (as well as complexes $Ni_2(OAc)_3(acac)\times MP\times2H2O$) this structures organization is not observed. It was established that these iron constructions are not formed in the absence of the aqueous environment. As mentioned above we showed the participation of H_2O molecules in mechanism of $Fe^{III,II}_x(acac)_y18C6_m(H_2O)_n$ transformation by analogy Dkel action, and also the increase in catalytic activity of iron complexes ($Fe^{III}_x(acac)$ $_y18C6_m(H_2O)_n$, $Fe^{II}_x(acac)_y18C6_m(H_2O)_n$ and $Fe^{II}_xL^1_y(L^1_{ox})_z(18C6)_n(H_2O)_m$) in the ethyl benzene oxidation in the presence of small amounts of water [1,2, 27]. After our works in article [65] it was found that the possibility of decomposition of the b-diketone in iron complex by analogy with Fe-ARD' action increases in aquatic environment. That apparently is consistent with data, obtained in our previous works [1, 2, 27].

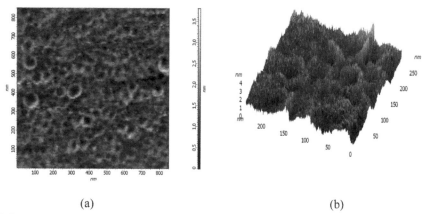

(a)   (b)

FIGURE 8.8   The AFM two- (a) and three-dimensional (b) image of nanoparticles on the basis $Fe_x(acac)_y18C6_m(H_2O)_n$ formed on the surface of modified silicone.

As another example we researched the possibility of the supramolecular nanostructures formation on the basis of $Fe_x(acac)_yCTAB_m$ at putting of solutions of $Fe_x(acac)_yCTAB_m$ in $CHCl_3$ (Fig. 8.10) or H_2O (Fig. 8.9) on the hydrophobic surface of modified silicon (CTAB=Me_3(n-C16H33)NBr). We used CTAB concentration in the 5–10 times less than Fe(acac)$_3$ concentration to reduce the probability of formation of micelles in water. But formation of spherical micelles at these conditions cannot be excluded. Salts QX are known to form with metal compounds complexes of variable composition depending on the nature of the solvent [2]. So the formation of hetero-ligand complexes Fe$_x$(acac)$_y$CTAB$_m$(CHCl$_3$)$_p$ (and Fe$^{II}_x$(acac)$_y$(OAc)$_z$(CTAB)$_n$(CHCl$_3$)$_q$ also) seems to be probable [23].

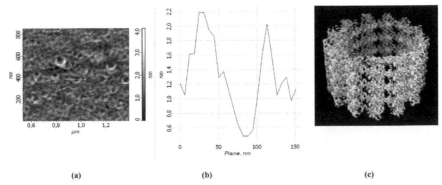

(a)   (b)   (c)

FIGURE 8.9   The AFM two-dimensional image (a) of nanoparticles on the basis Fe$_x$(acac)$_y$18C6$_m$(H$_2$O)$_n$ formed on the hydrophobic surface of modified silicone. The section of a circular shape with fixed length and orientation is about 50–80 nm (b), (c) The structure of the cell microtubules.

Earlier we established outer-sphere complex formation between Fe(acac)$_3$ and quaternary ammonium salt R$_4$NBr with different structure of R-cation [2]. Unlike the action of 18C6, in the presence of salts Me$_4$NBr, Me$_3$(n-C16H33) NBr (CTAB), Et$_4$NBr, Et$_3$PhNCl, Bu$_4$NI and Bu$_4$NBr, a decrease in the maximum absorption of acetylacetonate ion (acac)$^-$, and its bathochromic shift ($\Delta\lambda \approx 10$ nm) were observed (in CHCl$_3$). Such changes in the UV-spectrum reflect the effect of R$_4$NX on the conjugation in the (acac)$^-$ ligand in the case of the outer-sphere coordination of R$_4$NX. A change in the conjugation in the chelate ring of the acetylacetonate complex could be due to the involvement of oxygen atoms of the acac ligand in the formation of covalent bonds with the nitrogen atom or hydrogen bonds with CH groups of alkyl substituents [2].

In the Fig. 8.8, three-dimensional (a) and two-dimensional (b, d) AFM image of the structures on the basis of iron complex–Fe$_x$(acac)$_y$CTAB$_m$(CHCl$_3$)$_p$, formed at putting a uterine solution on a surface of modified silicone are presented.

As one can see, these nanostructures are similar to macrostructures, observed previously, but with less explicit structures, presented in Figs. 8.5 and 8.6, which resemble the shape of tubule micro fiber cavity (Fig. 8.9c). The height of particles on the basis of $Fe_x(acac)_y CTAB_m (CHCl_3)_p$ is about 7–8 nm (Fig. 8.10c), that more appropriate parameter for particle with $18C6\ Fe_x(acac)_y 18C6_m(H_2O)_n$. We showed that complexes of nickel $NiI^{I}(acac)_2\ CTAB$ (1:1) (in $CHCl_3$) do not form similar structures.

In the Fig. 8.9, two-dimensional (a–c) AFM image of the structures on the basis of iron complex with CTAB: $Fe_x(acac)_y CTAB_p(H_2O)_q$, formed at putting a uterine water solution on a hydrophobic surface of modified silicone, are presented. In this case, we observed apparently phenomenon of the particles, which are the remnants of the micelles formed. Obviously, the spherical micelles, based on $Fe(acac)_3$ and CTAB, on a hydrophobic surface are extremely unstable and decompose rapidly. The remains of the micelles have circular shaped structures, different from that observed in Figs. 8.6–8.8. The heights of particles on the basis of $Fe_x(acac)_y CTAB_p(H_2O)_q$ (Fig. 8.11) are about 12–13 nm that are greater than in the case of particles presented in Figs. 8.9 and 8.10. The particles in Fig. 8.11 resemble micelles in sizes (particle sizes in the XY plane are 100–120 nm (Fig. 8.11c)) and also like micelles, probably due to the rapid evaporation of water needed for their existence, are very unstable and destroyed rapidly (even during measurements).

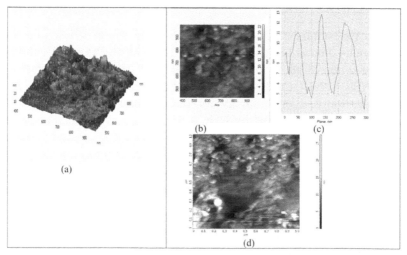

FIGURE 8.10   The AFM three- (a) two-dimensional image (b, d) of nanoparticles on the basis $Fe_x(acac)_{yCTABm}(CHCl_3)_p$ formed on the hydrophobic surface of modified silicone. The section of a circular shape with fixed length and orientation is about 50–70 nm (c).

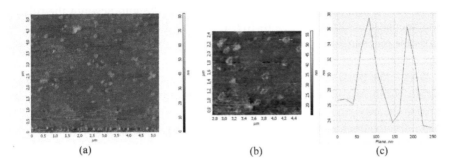

(a)   (b)   (c)

FIGURE 8.11 The AFM two-dimensional image (a, b) of nanoparticles (the remains of the micelles) on the base of $Fe_x(acac)_yCTAB_p(H_2O)_q$ formed on the hydrophobic surface of modified silicone. The section of a circular shape with fixed length and orientation is about 100–120 nm (c).

Unlike catalysis with iron-Dioxygenase, mechanism of catalysis by the Ni^{II}ARD does not include activation of O_2, and oxygenation of Acireductone leads to the formation of products not being precursors of methionine [62]. Earlier we have showed that formation of multidimensional forms based on nickel complexes can be one of the ways of regulating the activity of two enzymes [64]. The association of $Ni_2(AcO)_3(acac)\times MP\times 2H_2O$, which is functional and structure model of Ni-ARD, to supramolecular nanostructure due to intermolecular H-bonds (H_2O–MP, H_2O–(OAc$^-$) (or (acac$^-$)), is demonstrated on the next (Fig. 8.12). All structures (Fig. 8.12) are various on heights from the minimal 3–4 nm to ~20–25 nm for maximal values (in the form reminding three almost merged spheres) [64].

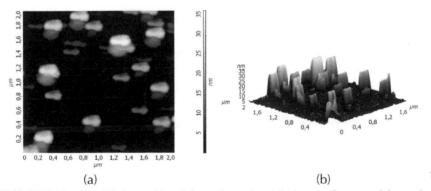

(a)   (b)

FIGURE 8.12 The AFM two- (a) and three-dimensional (b) image of nanoparticles on the basis $Ni_2(AcO)_3(acac)\times L^2\times 2H_2O$ formed on the hydrophobic surface of modified silicone.

Hear we assume that it may be necessary to take into account the role of the second coordination sphere [62, 64], including Tyr-fragment as one of possible mechanisms of reduce in enzymes activity in $Ni^{II}(Fe^{II})$ARD enzymes operation (the so-called mechanism of "switching functionality").

As it has been shown in Ref. [66], Tyr-fragment may be involved in substrate H-binding in step of O_2-activation by iron catalyst, and this can decrease the oxygenation rate of the substrate, as it is assumed in the case of Homoprotocatechuate 2,3-Dioxygenase. Tyr-fragment is discussed as important in methyl group transfer from S-adenosylmethionine (AdoMet) to dopamine [67]. The experimental findings with the model of Methyltransferase and structural survey imply that methyl CH×O hydrogen bonding (with participation of Tyr-fragment) represents a convergent evolutionary feature of AdoMet-dependent Methyltransferases, mediating a universal mechanism for methyl transfer [68].

In the case of Ni-Dioxygenase ARD, Tyr-fragment, involved in the mechanism, can reduce the Ni^{II}ARD-activity. Really, we have found earlier [2] that the inclusion of PhOH in complex $Ni(acac)_2 \cdot L^2$ (L^2=N-methylpyrrolidone-2), which is the primary model of Ni^{II}ARD, leads to the stabilization of formed triple complex $Ni(acac)_2 \times L^2 \times PhOH$. In this case, as we have established, ligand (acac)$^-$ is not oxygenated with molecular O_2. Also the stability of triple complexes $Ni(acac)_2 \times L^2 \times PhOH$ seems to be due to the formation of stable to oxidation of supramolecular macrostructures due to intra and intermolecular H-bonds. Formation of supramolecular macrostructures due to intermolecular (phenol–carboxylate) H-bonds and, possible, the other noncovalent interactions [69], based on the triple complexes $Ni(acac)_2 \cdot L^2 \cdot PhOH$, established by us with the AFM-method [70–72] (in the case of L^2=NaSt, LiSt) (Fig. 8.13a–c), is in favor of this hypothesis.

At the same time it is necessary to mean that important function of Ni^{II}ARD in cells is established now. Namely, carbon monoxide, CO, is formed as a result of action of nickel-containing Dioxygenase Ni^{II}ARD. It was established, that CO is a representative of the new class of neural messengers, and seems to be a signal transducer like nitrogen oxide, NO [11, 62]

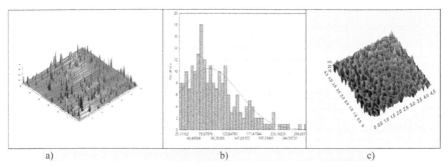

a)   b)   c)

FIGURE 8.13 (a). The AFM three-dimensional image (30×30(μm)) of the structures formed on a surface of modified silicone on the basis of triple complexes $Ni^{II}(acac)_2 \cdot NaSt \cdot PhOH$. (b) Histogram of mean values of height (~80 nm) of the AFM images of nanostructures based on $Ni^{II}(acac)_2 \cdot NaSt \cdot PhOH$, formed on the surface of modified silicone. (c) The AFM three-dimensional image (4.5×4.5(μm)) of the structures (h ~ 10 nm) formed on a surface of modified silicone on the basis of triple complexes $Ni^{II}(acac)_2 \cdot LiSt \cdot PhOH$.

8.7 CONCLUSIONS

Usually in the quest for axial modifying ligands that control the activity and selectivity of homogeneous metal complex catalysts, the attention of scientists is focused on their steric and electronic properties. The interactions in the outer coordination sphere, the role of hydrogen bonds and also the other noncovalent interactions are less studied.

In recent years, it was found, that water molecules present in a protein active site may play not only the structuring role but also serve as a reagent in biochemical processes.

In our works studying the effect of addition of small amounts of water to catalytic systems confirmed the important role of H-bonds in mechanisms of catalysis. It was found that the introduction of small portions of water to catalytic systems based on iron complexes $\{FeL^1_{n+L}{}^2\}$ (L^2 are crown ethers or quaternary ammonium salts) leads to the higher catalytic activity of these systems.

We have assumed that the high stability of heteroligand $M^{II}_{xL}{}^1_y(L^1_{ox})$ $_z(L^2)_n$ complexes as selective catalysts of the ethylbenzene oxidation to PEH, formed during the ethylbenzene oxidation in the presence of $\{ML^1_{n+}L^2\}$ systems as a result of oxygenation of the primary complexes $(M^{II}L^1_2)_x(L^2)_y$, can be associated with the formation of the supramolecular structures due to the intermolecular H-bonds.

The supramolecular nanostructures on the basis of iron $(Fe^{III}L^1_n)_x(L^2)_y$, $(L^1 = acac^-$, $L^2 = 18C6$, CTAB) and nickel $Ni^{II}_{xL}{}_y^1(L^1_{ox})_{z(L}{}^2)_n$ ($L^1 = acac^-$, $L^1_{ox = OAc}$, $L^2 = $ N-methylpirrolidone-2) complexes formed with assistance of intermolecular H-bonds, obtained with AFM method, indicate high probability of supramolecular structures formation due to H-bonds in the real systems, namely, in the processes of alkylarens oxidation. So the H-bonding seems to be one of the factors, responsible for the high activity and stability of catalytic systems researched by us.

Since the investigated complexes are structural and functional models of $Ni^{II}(Fe^{II})$ARD Dioxygenases, the data could be useful in the interpretation of the action of these enzymes.

Specific structural organization of iron complexes may facilitate the first step in Fe^{II}ARD operation: O_2 activation and following regioselective addition of activated oxygen to Acireductone ligand (unlike mechanism of regioselective addition of no activated O_2 to Acireductone ligand in the case of Ni^{II}ARD action), and reactions leading to formation of methionine The formation of multidimensional forms (in the case of Ni^{II}ARD) may be one way of controlling $Ni^{II}(Fe^{II})$ARD activity. The role of the second coordination sphere in mechanism of "switching functionality" in $Ni^{II}(Fe^{II})$ARD operation, is discussed.

KEYWORDS

- a-phenyl ethyl hydroperoxide
- Additives of small amounts of H_2O
- AFM method
- Ammonium quaternary salts
- Dioxygen
- Ethylbenzene
- $Fe^{II,III}$-acetylacetonates
- $Fe_x(acac)_yCTAB_m$
- Homogeneous catalysis
- Macro-cycle polyethers
- Nanostructures based on $Fe^{III}_x(acac)_y18C6_m(H2O)_n$
- $NixL^1y(L^1ox)_z(L^2)_n(H_2O)_m$
- Oxidation

REFERENCES

1. Matienko, L. I. (2007). Solution of the Problem of Selective Oxidation of Alkylarenes by Molecular Oxygen to Corresponding Hydro Peroxides, Catalysis Initiated by Ni(II), Co(II), and Fe(III) Complexes Activated by Additives of Electron-Donor Mono or Multi Dentate Extra-Ligands, *In Reactions and Properties of Monomers and Polymers* (Amore, A. D., Zaikov, G. Eds.) (New York: Nova Science Publishers), 21–41.
2. Matienko, L. I., Mosolova, L. A., & Zaikov, G. E. (2010). Selective Catalytic Hydrocarbons Oxidation, New Perspectives, New York, Nova Science Publ. Inc., USA, 150p.
3. Borovik, A. S. (2005). Bio Inspired Hydrogen Bond Motifs in Ligand Design, The Role of Non Covalent Interactions in Metal Ion Mediated Activation of Di oxygen, Acc. Chem. Res., *38*, 54–61.
4. Holm, R. H., & Solomon, E. I. (2004). Bio mimetic Inorganic Chemistry, Chem. Rev., *104*, 347–348.
5. Tomchick, D. R., Phan, P., Cymborovski, M., Minor, W., & Holm, T. R. (2001). Structural and Functional Characterization of Second-Coordination Sphere Mutants of Soybean Lipoxygenase-1, Biochemistry, *40*, 7509–7517.
6. Perutz, M. F., Fermi, G., Luisi, B., Shaanan B., & Liddington, R. C. (1987). Stereochemistry of Cooperative Mechanisms in Hemoglobin, Acc. Chem. Res., *20*, 309–321.
7. Schlichting, I., Berendzen, J., Chu, K., Stock, A. M., Maves, S. A., Benson, D. E., Sweet, R. M., Ringe, D., Petsko, G. A., & Sligar, S. G. (2000). The Catalytic Pathway of Cytochrome P450cam at Atomic Resolution, Science, *287*, 1615–1622.
8. Uehara, K., Ohashi, Y., & Tanaka, M. (1976). Bis(acetylacetonato) Metal (II)–Catalyzed Addition of Acceptor Molecules to Acetylacetone, Bull. Chem. Soc. Japan, *49*, 1447–1448.
9. Robie Lucas, L., Matthew Zart, K., Jhumpa Murkerjee, Thomas Sorrell, N., Douglas Powell, R., & Borovik, A. S. (2006). A Modular Approach toward Regulating the Secondary Coordination Sphere of Metal Ions, Differential Di oxygen Activation Assisted by Intra molecular Hydrogen Bonds, *J. Am. Chem. Soc, 128*, 15476–15489.
10. Nelson, J. H., Howels, P. N., Landen, G. L., De Lullo, G. S., & Henry, R. A. (1979). Catalytic Addition of Electrophiles to β-dicarbonyls, *In* Fundamental Research in Homogeneous Catalysis, *3*, (New York, London Plenum,), 921–939.
11. Dai, Y., Th Pochapsky, C., & Abeles, R. H. (2001). Mechanistic Studies of Two Di Oxygenases in the Methionine Salvage Pathway of Klebsiella Pneumonia, *Biochemistry, 40*, 6379–6387.
12. Gopal, B., Madan, L. L., Betz, S. F., & Kossiakoff, A. A. (2005). The Crystal Structure of a Quercetin 2, 3-Dioxygenase from Bacillus Subtilis Suggests Modulation of Enzyme Activity by a Change in the Metal Ion at the Active Site(s), *Biochemistry, 44*, 193–201.
13. Balogh-Hergovich, E., Kaizer, J., & Speier, G. (2000). Kinetics and Mechanism of the Cu (I) and Cu(II) Flavonolate-Catalyzed Oxygenation of Flavonols, Functional Quercetin 2, 3-dioxygenase Models, *J. Mol. Catal. A Chem., 159*, 215–224.
14. Straganz, G. D., & Nidetzky, B. (2005). Reaction Coordinate Analysis for β-Diketone Cleavage by the Non-Heme Fe^{2+}-Dependent Dioxygenase Dke 1, *J. Am. Chem. Soc., 127*, 12306–12314.
15. Klibanov, A . M . (1990). Asymmetric Transformations Catalyzed by Enzymes in Organic Solvents, Acc. Chem. Res., *23*, 114–120.
16. Ball, Ph. (2008). Water as an Active Constituent in Cell Biology, Chem. Rev., 108, 74.
17. Derat, E., Shaik, S., Rovira, C., Vidossich, P., & Alfonso-Prieto, M. (2007). The Effect of a Water Molecule on the Mechanism of Formation of Compound 0 in Horseradish Peroxidase, J. Am. Chem. Soc., *129*, 6346–6347.

18. Ma, L. H., Liu, Y., Zhang, X., Yoshida, T., Langry, K. C., Smith, K. M., & LaMar, G. N. (2006). Modulation of the Axial Water Hydrogen-Bonding Properties by Chemical Modification of the Substrate in Resting State, Substrate-Bound Heme Oxygenase from *Neisseria Meningitides*, Coupling to the Distal H-Bond Network via Ordered Water Molecules, J. Am. Chem. Soc., *128,* 6391–6399.

19. Matsui, T., Omori, K., Jin, H., & Ikeda-Saito, M. (2008). Alkyl Peroxides Reveal the Ring Opening Mechanism of Verdoheme Catalyzed by Heme Oxygenase, J. Am. Chem. Soc., *130,* 4220–4221.

20. Zotova, N., Franzke, A., Armstrong, A., & Blackmond, D. G. (2007). Clarification of the Role of Water in Proline-Mediated Aldol Reactions, J. Am. Chem. Soc., *129,* 15100–15101.

21. Piera, J., Persson, A., Caldentey, X., & Backvall, J. E. (2007). Water as Nucleophile in Palladium-Catalyzed Oxidative Carbohydroxylation of Allene-Substituted Conjugated Dienes, J. Am. Chem. Soc., *129,* 14120–14121.

22. Funabiki, T., Yoneda, I., Ishikawa, M., Ujiie, M., Nagai, Y., & Yoshida, S. (1994). Extradiol Oxygenation of 3, 5-Di-tert-butylcatechol with O_2 by Iron Chlorides in Tetrahydrofuran Water as a Model Reaction for Catechol 2, 3-dioxygenases, J. Chem. Soc., Chem. Commun., 1453–1454.

23. Nekipelov, V. M., & Zamaraev, K. I. (1985). Outer-Sphere Coordination of Organic Molecules to Electric Neutral Metal Complexes, Coord Chem. Rev., *61,* 185–240.

24. Pardo, L., Osman, R., Weinstein, H., & Rabinowitz, J. R. (1993). Mechanisms of Nucleophilic Addition to Activated Double Bonds, 1, 2 and 1, 4 Michael Addition of Ammonia, J. Am. Chem. Soc., *115,* 8263–8269.

25. Csanyi, J., Jaky, K., Dombi, G., Evanics, F., Dezso, G., & Kota, Z. (2003) Onium-Decavanadate Ion-Pair Complexes as Catalysts in the Oxidation of Hydrocarbons by O_2, J. Mol. Catal A: Chem., *195,* 101–110.

26. Csanyi, L. J., Jaky, K., Palinko, I., Lockenbauer, A., & Korecz, L. (2000). The Role of Onium Salts in the Oxidation of Hydrocarbons by O_2 Catalyzed by Cationic Phase-Transfer Reagents, Phys. Chem. Chem. Phys, 2, 3801–3805.

27. Matienko, L. I., & Mosolova, L. A. (2008). Modeling of Catalytic of Complexes $Fe^{II,III}(acac)_n$ with R_4NBr or 18-Crown-6 in the Ethyl benzene Oxidation by Dioxygen in the Presence of Small Amounts of H_2O, Oxid Commun., *31,* 285–298.

28. Matienko, L. I., & Mosolova, L. A. (2008b). The Effect of Small Amounts of H_2O on Mechanism of the Ethyl benzene Oxidation by Dioxygen, Catalyzed with Complexes $Fe^{II,III}(acac)_n$ with 18-Crown-6, Neftekhimiya, *48,* 368–377.

29. Dehmlow, E., & Dehmlow, Z. (1987). Phase Transfer Catalysis (M: Mir Press,) 485P.

30. Matienko, L. I., & Mosolova, L. A. (2007). Selective Oxidation of Ethyl Benzene with Dioxygen into a-Phenylethylhydroperoxide, Modification of Catalyst Activity of Ni (II) and Fe(III) complexes Upon Addition of Quaternary Ammonium Salts as Exoligands, *In* New Aspects of Biochemical Physics. Pure and Applied Sciences (Eds. Varfolomeev, S. D., Burlakova, E. B., Popov, A. A., Zaikov, G. E.) (New York: Nova Science Publishers), 95–112.

31. Emanuel, N. M., Zaikov, G. E., & Maizus, Z. K. (1973). Rol' Sredy v Radikal'no-tsepnykh Reaktsiyakh Okisleniya Organicheskikh Soedinenii (Role of the Medium in Chain-radical Oxidation of Organic Compounds) (Moscow, Nauka)

32. Partenheimer W. (1991). Metodology and Scope of Metal/Bromide Autoxidation of Hydrocarbon Catal Today, *23,* 69–158.

33. Panicheva, L. P., Turnaeva, E. A., Panichev, S. A., Ya Yuffa, A. (1994b). The Characteristics of Liquid-Phase Oxidation of Cumene with Molecular Oxygen in the Presence of Cu(II) Inverted Micelles Systems, *Neftekhimiya, 34,* 171–179.

34. Maksimova, T. V., Sirota, T. V., Koverzanov, E. V., & Kasaikina, O. T. (2001b). The Influence of Surfactants on the Oxidation of Ethyl benzene, 4, Catalysis of Oxidation of Ethyl Benzene with Cetyltrimethyl Ammonium Bromide in Combination with Cobalt(II) Acetylacetonate, Neftekhimiya, *41*, 289–293.

35. Matienko, L. I., & Mosolova, L. A. (2006). Mechanism of Selective Oxidation of Ethyl Benzene with Dioxygen Catalyzed by Complexes of Nickel (II), Cobalt (II) and Iron (III) with Macro Cyclic Poly Ethers, *In* Uspekhi v Oblasti Geterogennogo Kataliza i Geterotsiklov (Progress in Heterogeneous Catalysis and Heterocycles) (Rakhmankulov, D. L. Ed.) (Moscow, Khimiya,), 235–260.

36. Belsky, V. K., & Bulychev, B. M. (1999). Structurally Chemical Aspects of Complex Forming in Systems Metal Halide, Macrocyclic Polyether, *Usp Khim.*, *68*, 136–153 [Russ Chem Rev., *68*, 119–135].

37. Kuramshin, E. M., Kochinashvili, M. V., Zlotskii, S. S., & Rakhmankulov, D. P. (1984f). Influence of Crown-Ethers on Homogeneous-Catalytic Oxidation of 1, 3-dioxolan, Dokl. Akad. Nauk SSSR, 279, 1392–1397.

38. Goldberg, I. (1978). Study of Host Guest Complexation of Alkalin Earth Metal Ion, Aluminium Ion and Transition Metal Ions with Crown Ethers by Fast Atom Bombardment Mass Spectrometry, *Acta Cryst, 34,* 3387–3394.

39. Jaguar-Grodzinnski, J. (1985). Protonation of Macrocyclic Polyethers, Isr *J. Chem., 25,* 39.

40. Mosolova, L. A., Matienko, L. I., & Skibida, I. P. (1994). Composition Catalysts of Ethyl Benzene Oxidation Based on Bis (Acetylacetonato) Nickel (II) and Phase Transfer Catalysts as Ligands, 1 *Macro Cyclic Poly Ethers,* Izv. AN, Ser. Khim, 1406–1411 (in Russian).

41. Mosolova, L. A., Matienko, L. I., & Skibida, I. P. (1994). Composition Catalysts of Ethyl Benzene Oxidation Based on Bis(Acetylacetonato) Nickel (II) and Phase Transfer Catalysts as Ligands, 2 *Quarternary Ammonium Salts,* Izv. AN, Ser. Khim, 1412–1417 (in Russian).

42. Matienko, L. I., & Mosolova, L. A. (2007). Selective Oxidation of Ethyl Benzene with Dioxygen into a-Phenyl ethyl hydro peroxide, Modification of Catalyst Activity of Ni (II) and Fe (III) Complexes Upon Addition of Quaternary Ammonium Salts as Exo ligands, *In New Aspects of Bio Chemical Physics Pure and Applied Sciences,* Varfolomeev, S. D., Burlakova, E. B., Popov, A. A., Zaikov, G. E. Eds. (New York: Nova Science Publishers), 95–120.

43. Denisov, E. T., & Emanuel, N. M. (1960). Catalysis with Salts of Transition Metals in Reactions of Liquid Phase Oxidations, *Uspekhi khimii, 29,* 1409–1420 [Russ. Chem. Rev., 29, 645–656].

44. Emanuel, N. M., & Gall, D. (1984). Okislenie Etilbenzola Model'naya Reaktsiya (Oxidation of Ethyl Benzene Model Reaction) (Moscow Nauka).

45. Chaves, F. A., Rowland, J. M., Olmstead, M. M., & Mascharak, P. K. (1998). Syntheses Structures and Reactivates of Cobalt(III)-Alkyl Peroxo Complexes and their Role in Stoichiometric and Catalytic Oxidation of hydrocarbon, *J. Am. Chem. Soc., 120,* 9015–9025.

46. Solomon-Rapaport, E., Masarwa, A., Cohen, H., & Meyerstein, D. (2000). On the Chemical Properties of the Transient Complexes L_mM^{n+1}-OOR, Part I. The Case of (2, 3, 9, 10-tetramethyl 1, 4, 8, 11-Tetraaza Cyclotetradeca 1, 3, 8, 10-tetraene) (H_2O) Co^{III}-$OOCH_3^{2+}$, In org. Chim. Acta, *299,* 41–46.

47. Semenchenko, A. E., Solyanikov, V. M., & Denisov, E. T. (1973). Reactions of Cyclohexyl Peroxo Radicals with Stearates of Transition Metals, Zh Phiz Khimii, *47,* 1148, (*in Russian*) d

48. Morassi, R., Bertini, I., & Sacconi, L. (1973). Five-Coordination in Fe (II), Co (II) and Ni (II) Complexes, Coord. Chem. Rev., *11,* 343–402.

49. Alekseevski, V. A. (1982). Research of Complex Formation of 3d-metal β-diketonates with Amines, In Problemy Khimii i Primeneniya Diketonatov Metallov (The Problems of the Chemistry and Applications of Metal β-Diketonates). Spizin, V. I. Ed., M. "Nauka," 65–71.

50. Stynes, D. V., Stynes, H. C., James, B. R., & Ibers, J. A. (1973). Thermodynamics of Ligand and Oxygen Binding to Cobalt Protoporphyrin IX Dimethyl Ester in Toluene Solution, J. Am. Chem. Soc., 95, 1796–1801.

51. Dedieu, A., Rohmer, M. M., Benard, M., & Veillard, A. (1976). Oxygen Binding to Iron Porphyrins, an Ab Initio Calculation, J. Am. Chem. Soc., 98, 3717–3718.

52. Komiya, N., Naota, T., & Murahashi, S. I. (1996). Aerobic Oxidation of Alkanes in the Presence of Acetaldehyde Catalyzed by Copper-Crown Ether, Tetrahedron Lett, 37, 1633–1636.

53. Ch Kang, K., Redman, Ch., Cepak, V., & Sawyer, D. T. (1991). Iron (II)/ Hydro Peroxide (Fenton Reagent) Induced Activation of Dioxygen for (A) the Direct Ketonization of Methylenic Carbon and (B) the Dioxygenation of Cis-Stilbene, Bioorg. Med. Chem., 1, 125.

54. Matsushita, T., Sawyer, D. T., & Sobkowiak, A. (1999). Mn (III) L_x/t-BuOOH Induced Activation of Di oxygen for the Di Oxygenation of Cyclohexane, J. Mol. Catal A Chem, 137, 127.

55. Saggu, M., Levinson, N. M., & Boxer, S. G. (2011). Direct Measurements of Electric Fields in Weak OH×π Hydrogen Bonds, J. Am. Chem. Soc., 133, 17414–17419.

56. Ma, J. C., & Dougherty, D. A. (1997). The Cation−π Interaction, Chem. Rev., 97, 1303–1324.

57. Graham, J. D., Buytendyk, A. M., Wang, Di, Bowen, K. H., & Collins, K. D. (2014). Strong Low-Barrier Hydrogen Bonds May Be Available to Enzymes, Biochemistry, 53, 344–349.

58. Leninger, St., Olenyuk, B., & Stang, P. J. (2000). Self-Assembly of Discrete Cyclic Nanostructures Mediated by Transition Metals, Chem. Rev., 100, 853–908.

59. Stang, P. J., & Olenyuk, B. (1997). Self-Assembly, Symmetry, and Molecular Architecture: Coordination as the Motif in the Rational Design of Supra Molecular Metallacyclic Polygons and Polyhedra, Acc. Chem. Res., 30, 502–518.

60. Drain, C. M., & Varotto Radivojevic, A. I. (2009). Self-Organized Porphyrinic Materials, Chem. Rev., 109, 1630–1658.

61. Beletskaya, I., Tyurin, V. S., Tsivadze Yu, A., & Guilard, R. (2009). Ch. Stem Supra Molecular Chemistry of Metalloporphyrins, Chem. Rev., 109, 1659–1713.

62. Chai, S. C., Ju, T., Dang, M., Goldsmith, R. B., Maroney, M. J., & Pochapsky Th C. (2008). Characterization of Metal Binding in the Active Sites of Acireductone Dioxygenase Isoforms from Klebsiella ATCC 8724, Biochemistry, 47, 2428–2435.

63. Leitgeb, St., Straganz, G. D., & Nidetzky, B. (2009). Functional Characterization of an Orphan Cupin Protein from Burkholderia Xenovorans Reveals A Mononuclear Nonheme Fe^{2+} Dependent Oxygenase that Cleaves b-diketones, The FEBS Journal, 276, 5983–5997.

64. Matienko, L. I., Mosolova, L. A., Binyukov, V. I., Mil, E. M. & Zaikov, G. E. (2012). "The New Approach to Research of Mechanism Catalysis with Nickel Complexes in Alkylarens Oxidation" "Polymer Yearbook" (2011). New York: Nova Science Publishers, 221–230.

65. Allpress, C. J., Grubel, K., Szajna-Fuller, E., Arif, A. M., & Berreau, L. M. (2013). Regioselective Aliphatic Carbon-Carbon Bond Cleavage by Model System of Relevance to Iron-Containing Acireductone Dioxygenase, J.Am.Chem.Soc, 135, 659–668.

66. Mbughuni, M. M., Meier, K. K., Münck, E., & Lipscomb, J. D. (2012). Substrate-Mediated Oxygen Activation by Homoprotocatechuate 2, 3-Dioxygenase, Intermediates formed by a Tyrosine257 Variant, Biochemistry, 51, 8743.

67. Zhang, J., & Klinman, J. P. (2011). Enzymatic Methyl Transfer, Role of an Active Site Residue in Generating Active Site Compaction that Correlates with Catalytic Efficiency, J. Am. Chem. Soc., 133, 17134–17137.

68. Horowitz, S., Dirk, L. M. A., Yesselman, J. D., Nimtz, J. S., Adhikari, U., Mehl, R. A., St. Scheiner, R., Houtz, L., Al-Hashimi, H. M., & Trievel, R. C. (2013). Conservation and Functional Importance of Carbon-Oxygen Hydrogen Bonding in Ado Met-Dependent Methyl transferases, *J. Am. Chem. Soc., 135,* 15536–15548.

69. Mukherjee, P., Drew, M. G. B., Gómez-Garcia, C. J., & Ghosh, A. (2009). (Ni_2), (Ni_3), and ($Ni_2 + Ni_3$) a Unique Example of Isolated and Cocrystallized Ni_2 and Ni_3 Complexes, Inorg. Chem., *48,* 4817–4825.

70. Matienko, L. I., Binyukov, V. I., Mosolova, L. A., Mil, E. M., & Zaikov, G. E. (2012). Supra Molecular Nanostructures on the Basis of Catalytic Active Hetero Ligand Nickel Complexes and their Possible Roles in Chemical and Biological Systems, *J. Biol. Research, 1,* 37–44.

71. Matienko, L. I., Binyukov, V. I., Mosolova, L. A., Mil, E. M. L., & Zaikov, G. E. (2014). The Supra Molecular Nanostructures as Effective Catalysts for the Oxidation of Hydrocarbons and Functional Models of Dioxygenases, *Polymers Research Journal, 8,* 91–116.

72. Ludmila Matienko, Vladimir Binyukov, Larisa Mosolova, & Gennady Zaikov (2011). The Selective Ethylbenzene Oxidation by Dioxygen into a-Phenyl Ethyl Hydroperoxide, Catalyzed with Triple Catalytic System $\{Ni^{II}(acac)_2 + NaSt(LiSt) + PhOH\}$, Formation of Nanostructures $\{Ni^{II}(acac)_2 \times NaSt \times (PhOH)\}_n$ with Assistance of Intermolecular H-bonds, *Polymers Research Journal, 5,* 423–431.

CHAPTER 9

NANOCOMPOSITE FOILS BASED ON SILICATE PRECURSOR WITH SURFACE-ACTIVE COMPOUNDS: A RESEARCH NOTE

L. O. ZASKOKINA, V. V. OSIPOVA, Y. G. GALYAMETDINOV, and A. A. MOMZYAKOV

CONTENTS

ABSTRACT

Obtained film composites containing complex of Eu (DBM) 3 bpy. It is shown that the using of liquid-crystal matrix results to a more uniform and orderly arrangement of the complex and silicate substrate.

9.1   INTRODUCTION

Liquid crystals are extremely flexible that is driven under the influence of a relatively weak external factors leading to changes in macroscopic physical properties of the sample. Therefore, materials based on them have found practical application in the most advanced fields of science and technology [1].

Great interest is the creation based on lyotropic liquid crystals of nanoscale materials with improved physical properties by doping silicate matrices and lanthanide ions and other metals into them [2, 3]. Adding silicate matrix to LCD systems allows stabilizing nano organized structure [4].

One of the most important areas of modern material engineering aligned with the obtaining of nanostructures with desired properties. For this widely used approach of obtaining composite nanomaterials, that is, particles of prisoners in a chemically inert matrix. In many cases, by way of such matrices we use different porous materials In these pores, various compounds may be administered and then, after chemical modification provide particles of the desired material, size and shape are the same shape as the cavities of the matrix, and its walls prevent their aggregation and to protect against environmental influences [5]. The aim of this work was to obtain nano composites lanthanide complexes with silicate matrices.

9.2   EXPERIMENTAL PART

Silicate matrices were prepared by the sol-gel process from reaction mixtures consisting of vinyl trimethoxysilane (VMOS)/phenyl trimethoxysilane (PhTEOS), ethanol and water [6]. Firstly, we measured the weight of precursor and then we added ethanol plus water. The samples obtained were placed on a magnetic stirrer at a rate of 450 rev/min at temperature of 50 °C for time duration of 2–4 days. Periodically we checked pH value (2, 3 times) time by time the water phase was transformed into a gel. A few hours later, we could observe a condensation into a solid continuous network. Cooked systems are characterized by composition of the initial components, that is the molar and weight ratio.

Lyomesophases synthesis was carried out as described in article [8], based on nonionic surfactants (C12EO4, C12EO9, C12EO10). Identification of liquid-crystalline properties conducted according to the polarization-optical microscopy (POM) (Olympus BX 51) the observed textures set the type of mesophase and the phase transition temperature.

Silicate was built in liquid crystal matrix between mi cellar aggregates. Incorporation of small amounts of hydrochloric acid necessary sought pH (2 ÷ 3) system, wherein the water phase is gelled. Few hours later we could observe the condensation into solid continuous network [9].

Using spin coating (SPIN COATER LAU-TELL WS-400-6NPP-LITE) we obtained films of nano composites including Si/C12EOn/H_2O. Following this method, we should introduce a system for glass and rotate it into device at a rate of 1000 rev/min.

Luminescent characteristics of multi component films by focusing mesophases become better, which allows passing from the supra molecular organization of the sample to streamline the entire volume.

9.3   IMPLICATIONS AND DISCUSSION

Control of the LCD systems synthesis process completeness performed by fixing constant temperature of mesophases transition, as a result, we got isotropic liquid in the whole volume of the sample.

Exploring LCD system textures, we found out that one system has a lamellar mesophases (C12EO14/H_2O), and two others have hexagonal one (C12EO10/H_2O, C12EO9/H_2O).

Systems were obtained with silicate matrix based on two different pre cursors vinyl trimethoxysilane and phenyl triethoxysilane by sol-gel technology. During the synthesis of the silicate matrix, molar ratios, aging time and drying time were the most important characteristics, so they were experimentally optimized (Table 9.1).

TABLE 9.1   Controlled Parameters during the Sol-Gel Synthesis

Factor	Range*	Optimum
pH	1–7	2, 3
		If the pH is too high (>pH 3.5), adding some amount of water, initial solution is turbid enough and the gel time is rapid
Sol temperature	18–100 °C	50 °C
		Samples are thickened quickly (a few days)

TABLE 9.1   *(Continued)*

Factor	Range*	Optimum
Aging time	1–20 days	2 days
		Drying time and temperature are the most important parameters
Drying temperature	18–100 °C	18 °C
		All samples are dried at elevated temperatures, cracked
Drying time	1 hour–6 months	>45 days for 4 mL of solution, it takes a long drying time.

* Published data.

After the admission of silicate into LCD, synthesized systems were investigated by polarized light. As in the original liquid crystal systems, appropriate texture observed that to characterize the supra molecular organization of the molecules in the mesophases.

Nanocomposite films were prepared by the method of spin coating. This method allows control of the film thickness, but this is a uniform distribution of the sample on the substrate.

Annealed porous films placed in a solution of the complex tris [1,3-diphenyl-1,3-propanedione] – [2,2'-bipyridyl] Europium Eu (DBM) 3 bpy in toluene for adsorption equation for 24hours. Then the samples were washed in toluene to remove the surfactant and water out of the surface and dried to remove the solvent [10]. After the process, the systems under study were examined by fluorescent spectroscopy (Fig. 9.1).

Lack of main energy transition signal splitting looks as shown: 5D0 → 7F1, 5D0 → 7F2 and 5D0 → 7F3 and indicates a low symmetry of the ligand environment of ion Eu^{3+}, which in principle expected in lyotropic systems containing a large amount of solvent. Small ratio of the peak areas of transitions 5D0 → 7F2 and 5D0 → 7F1 equal to 5.9 indicates a weak energy transfer from the ligand to the ion.

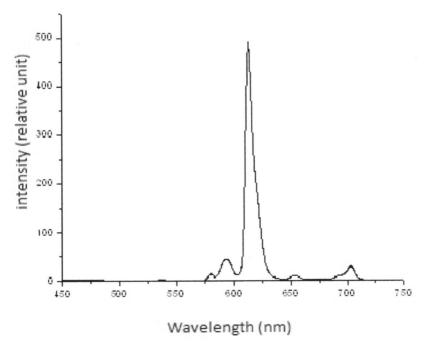

FIGURE 9.1 The luminescence spectrum and kinetics of luminescence system SiO2/ C12EO4/Eu (DBM) 3 bpy.

In films obtained on the glass, there is a uniform and orderly arrangement of the complex Eu (DBM) 3 bpy silicate and SiO$_2$, which increases the luminescence lifetime (Fig. 9.2).

Found out that oriented systems compared to disordered systems have more intense photoluminescence.

It should be emphasized the role of the orientation of mesophases as a necessary step in the organization of a liquid crystal template. Orientation mesophases switches from supramolecular organization in the domains of the sample to streamline the entire volume, which is especially important for the practical design of new functional materials when used in molecular electronics and laser optics.

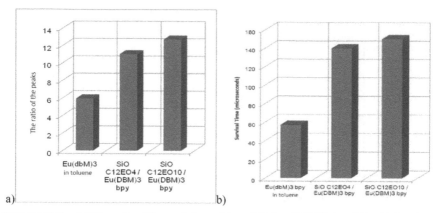

FIGURE 9.2   (a) The ratio of the peak intensities of 5D0 → 7F2/5D0 → 7F1; (b) Survival Time the glow luminescent complexes.

9.4   CONCLUSION

We obtained multicomponent Si/C12EOn/H$_2$O systems and films containing complex of Eu (DBM) 3 bpy. In films, there is a uniform and orderly arrangement of Eu (DBM) 3 bpy silicate and SiO$_2$, which increases the intensity and luminescence lifetime.

The study was supported by RFBR under research project number 12-08-31405 mol_a.

KEYWORDS

- **Aging time of luminescence**
- **Composite films**
- **Liquid crystals**
- **Luminescence**
- **Multi component systems**
- **Nanocomposites**
- **Silicate matrix**

REFERENCES

1. Blackstock, J. J., Donley, C. L, Stickle, W. F., Ohlberg, D. A. A., Yang, J. J., Stewart, D. R., & Williams, R. S. (2008). *J. Am. Chem. Soc., 130(12)*, 4041–4047.

2. Koen Binnemans, Yury, G., Galyametdinov, Rik Van Deun, Duncan Bruce, W. (2000). *J. Am. Chem. Soc., 122,* 4335–4344.
3. Duncan Bruce, W. (2000). Acc. Chem. Res., *33,* 831–840.
4. Ohtake, T., Ogasawara, M., Ito-Akita, K., Nishina, N., Ujiie, S., Ohno, H., & Kato, T. (2000). *Chem. Mater. 12,* 782.
5. Eliseev, A. A., Kolesnik, I. V., Lukashin, A. V., & Tretyakov, Y. D. (2005). *Adv. Eng. Mater., 7*(4), 213–217.
6. Margolin, V. I., Zharbeev, V. A., & Tupik, V. (2008). A Physical Basis of Microelectronics of Academy, 400p.
7. Osipova, V. V., Zaskokina, L. O., Gumerov, F. M., & Galyametdinov, J. G. (2012). Vestnik of KNRTU, *17,* 29–30.
8. Osipova, V. V., Selivanov, N. M., Danovski, D. E., & Galyametdinov, J. G. (2007). Vestnik of KNRTU, *6,* 30–35.
9. Osipova, V. V., Yarullin, L. Y., Gumerov, F. M., & Galyametdinov, J. G. (2010). Vestnik of KNRTU, *9,* 879–881.
10. Ostapenko, N. I., Kozlova, N. V., Frolov, E., Ostapenko, V., Pekus, D., Gulbinas, V., Eremenko, A. M., Smirnov, N., Surovtceva, N. I., Suto, S., & Watanabe, A. (2011). *Journal of Applied Spectroscopy, 78(1),* 82–88.

CHAPTER 10

INTERACTION OF THE 1,1,3-TRIMETHYL-3-(4-METHYLPHENYL) BUTYL HYDROPEROXIDE WITH TETRAETHYLAMMONIUM BROMIDE

N. A. TUROVSKIJ, YU. V. BERESTNEVA, E. V. RAKSHA,
M. YU. ZUBRITSKIJ, and G. E. ZAIKOV

CONTENTS

ABSTRACT

Kinetics of the activated 1,1,3-trimethyl-3-(4-methylphenyl) butyl hydro peroxide decomposition in the presence of tetra ethyl ammonium bromide (Et_4NBr) has been investigated. Et_4NBr has been shown to reveal the catalytic properties in this reaction. The complex formation between hydro peroxides and Et_4NBr has been shown by the kinetic and NMR ^{1}H spectroscopy methods. Thermodynamic parameters of the complex formation and kinetic parameters of complex-bonded hydro peroxide decomposition have been estimated. The structural model of the reactive hydro peroxide catalyst complex has been proposed. Complex formation is accompanied with hydro peroxide chemical activation.

10.1   INTRODUCTION

A promising and the easiest way of oxygen-containing compounds (alcohols, ketones, carboxylic acids, aldehydes, epoxides) producing, considering the requirements of "green chemistry" are the oxidation of hydrocarbons by molecular oxygen. Reactions of hydroperoxide compounds formation and decomposition are key to this process.

Quaternary ammonium salts exhibit a catalytic action for the reaction of radical decomposition of hydroperoxides [1, 2], that promotes the liquid-phase oxidation process of hydrocarbons by molecular oxygen under mild conditions [3, 4]. Variation of hydroperoxides and onium salts structure allows creating purposefully new initiator systems with predetermined reactivity. The results of kinetic studies showed that the catalysis of hydroperoxide decomposition in a presence of onium salts includes a stage of intermediate complex formation [1, 2, 4]. Different structural models of the hydroperoxide onium salt complexes were proposed on the base of molecular modeling results [2, 4].

The purpose of this work is a complex study of the 1,1,3-trimethyl-3-(4-methylphenyl) butyl hydroperoxide interaction with tetraethylammonium bromide by the kinetic, NMR ^{1}H spectroscopic and molecular modeling methods.

10.2   EXPERIMENTAL PART

1,1,3-Tri methyl-3-(4-methyl phenyl)butyl hydro peroxide (ROOH) was purified according to Ref. [5]. Its purity (98.9%) was controlled by iodometry method. Tetra alkyl ammonium bromide (Et_4NBr) was recrystallized from acetonitrile solution by addition of diethyl ether excess. The salt purity (99.6%)

was determined by argentum metric titration with potentiometric fixation of the equivalent point. Tetra alkyl ammonium bromide was stored in box dried with P_2O_5. Acetonitrile (CH_3CN) was purified according to Ref. [6]. Its purity was controlled by electro conductivity χ value, which was within $(8.5 \pm 0.2) \times 10^{-6} W^{-1} cm^{-1}$ at 303K.

Reaction of the hydro peroxide catalytic decomposition was carried out in glass-soldered ampoules in argon atmosphere. To control the proceeding of hydro peroxide thermolysis and its decomposition in the presence of Et_4NBr the iodometric titration with potentiometric fixation of the equivalent point was used.

NMR^{1}H spectroscopy investigations of the hydro peroxide and hydro peroxide Et_4NBr solutions were carried out in acetonitrile-d$_3$ (D_3CCN). The NMR^{1}H spectra were recorded on a Bruker Avance 400 (400 MHz) using TMS as an internal standard. D_3CCN was SIGMA-ALDRICH reagent and was used without additional purification, but it has being stored above molecular sieves for three days.

1,1,3-Tri methyl-3-(4-methyl phenyl)butyl hydro peroxide (4-CH_3-C_6H_4-$C(CH_3)_2$-CH_2-$(CH_3)_2$C-O-OH) NMR ^{1}H (400 MHz, acetonitrile-d$_3$, 297 K, δ ppm, J/Hz): 0.86 (s, 6 H, -C(CH_3)_2OOH), 1.33 (s, 6 H, -C_6H_4C(CH_3)_2-), 2.03 (s, 2 H, -CH_2-), 2.28 (s, 3 H, CH_3-C_6H_4-), 7.10 (d, J = 8.0, 2 H, H-aryl), 7.29 (d, J = 8.0, 2 H, H-aryl), 8.49 (s, 1 H, -COOH).

Tetraethylammonium bromide (Et_4NBr) NMR ^{1}H (400 MHz, acetonitrile-d$_3$, 297 K, δ ppm, J/Hz): = 1.21 (t, J = 8.0, 12 H, -CH_3), 3.22 (q, J = 8.0, 8 H, -CH_2-).

Quantum chemical calculations of the equilibrium structures of hydro peroxide molecule and corresponding radicals as well as tetra ethyl ammonium bromide and ROOH-Et_4NBr complexes were carried out by AM1 semiempirical method implemented in MOPAC2012™ package [7]. The RHF method was applied to the calculation of the wave function. Optimization of hydro peroxides structure parameters were carried out by Eigenvector following procedure. The molecular geometry parameters were calculated with boundary gradient norm 0.01. The nature of the stationary points obtained was verified by calculating the vibration frequencies at the same level of theory. Solvent effect in calculations was considered in the COSMO [8] approximation.

10.3 RESULTS AND DISCUSSION

Kinetics of the 1,1,3-trimethyl-3-(4-methyl) phenyl butyl hydro peroxide interaction with tetra ethyl ammonium bromide

Kinetics of the aralkyl hydro peroxides thermal decomposition has been investigated [9] in acetonitrile. Thermolysis of the 1,1,3-trimethyl-3-(4-methyl) phenyl butyl hydro peroxide was observed at 408–438K with appreciable rate. Rate of the hydro peroxide decomposition (W_0^{term}) has been estimated at 383K on the base of thermal dependence of the thermolysis rate constant and was 4.0×10^{-8} mol·dm^{-3} s^{-1} for initial hydro peroxide concentration [ROOH]$_0$ = 5.13×10^{-2} mol·dm^{-3}.

Kinetics of the 1,1,3-trimethyl-3-(4-methyl) phenyl butyl hydro peroxide ([ROOH] 0=5.13×10^{-2} mol·dm^{-3}) in the presence of tetra ethyl ammonium bromide ([Et$_4$NBr] 0=5.05×10^{-3} mol·dm^{-3}) has been studied in acetonitrile at 383K. Rate of the catalytic hydro peroxide decomposition W_0^{cat} was 5.8×10^{-7} mol·dm^{-3}×s^{-1} and on the order above its thermolysis rate. The Et$_4$NBr concentration reducing was not observed in the system at investigated conditions.

Further kinetics an investigation of the hydroperoxide decomposition in the presence of Et4NBr has been performed under conditions of ammonium salts excess in the reaction mixture. Reaction was carried out at 373–393 K, hydro peroxide initial concentration was 5×10^{-3} mol·dm^{-3}, Et$_4$NBr concentration in the system was varied within 2×10^{-2}–1.2×10^{-1} mol·dm^{-3}. Hydro peroxides decomposition kinetics could be described as the first order one. The reaction was monitored up to 80% hydro peroxide conversion and the products did not affect the reaction proceeding as the kinetic curves anamorphous were linear in the corresponding first order coordinates.

The effective rate constant (k_{ef}, s^{-1}) of the hydro peroxide decomposition determined in this conditions was found to be independent on the hydro peroxide initial concentration within Ref. [ROOH]$_0$ = 1×10^{-3}–8×10^{-3} mol dm^{-3} at 383 K while Et$_4$NBr amount was kept constant (5×10^{-2} mol dm^{-3}) in these studies. This fact allows one to exclude the simultaneous hydro peroxides reactions in the system under consideration.

Non-linear k_{ef} dependences on Et$_4$NBr concentration at the constant hydro peroxide initial concentration are observed for the studied system (Fig. 10.1a). The nonlinear character of these dependences allows us to assume that an intermediate adduct between ROOH and Et$_4$NBr is formed in the reaction of the hydro peroxide decomposition in the presence of Et$_4$NBr. These facts confirm to the kinetic scheme of activated cumene hydro peroxide and hydroxycyclohexyl hydro peroxide decomposition that has been proposed *previously* [1, 4].

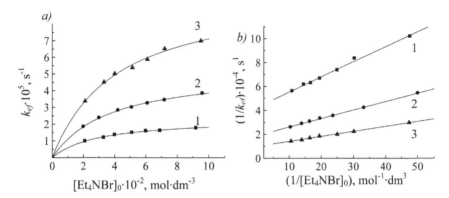

FIGURE 10.1 Dependence of k_{ef} on the Et$_4$NBr initial concentration in direct (*a*) and inverse (*b*) coordinates ([ROOH] 0=5.0×10^{-3} mol dm^{-3}, 383K) *1*–373 K; *2*–383 K; *3*–393K.

Chemically activated hydro peroxide decomposition in the presence of Et$_4$NBr is suggested to proceed in accordance with following kinetic scheme (Scheme 10.1). It includes the stage of a complex formation between the hydro peroxide molecule and Et$_4$NBr ions as well as the stage of complex bonded hydro peroxide decomposition.

$$\text{ROOH} + \text{Et}_4\text{NBr} \underset{}{\overset{K_C}{\rightleftharpoons}} [\text{complex}] \xrightarrow{k_d} \text{products} \qquad (1)$$

where K_C– *equilibrium* constant of the complex formation (dm^3·mol^{-1}), and k_d– rate constant of the complex decomposition (s^{-1}).

We assume the model of ROOH-Et$_4$NBr complex formation with combined action of cation and anion. The hydro peroxides thermolysis *contribution* to the overall reaction rate of the ROOH activated decomposition is negligibly small because thermolysis rate constants [9] are less by an order of magnitude than correspondent k_{ef} values. Using a kinetic model for the generation of active species (Scheme 10.1) and analyzing this scheme in a quasi-equilibrium approximation one can obtain the following equation for the k_{ef} dependence on Et$_4$NBr concentration:

$$k_{ef} = \frac{K_C k_d [\text{Et}_4\text{NBr}]_0}{1 + K_C [\text{Et}_4\text{NBr}]_0} \qquad (2)$$

To simplify further analysis of the data from Fig. 10.1a let us transform Eq. (2) into the following one:

$$\frac{1}{k_{ef}} = \frac{1}{k_d K_C [\text{Et}_4\text{NBr}]_0} + \frac{1}{k_d} \qquad (3)$$

Equation (3) can be considered as equation of straight line in $(\frac{1}{k_{ef}})$ – ($\frac{1}{[\text{Et}_4\text{NBr}]_0}$) coordinates. The k_{ef} dependences on Et_4NBr concentration are linear in double inverse coordinates (Fig. 10.1b). Thus the experimentally obtained parameters with reasonable accuracy correspond to the proposed kinetic model and are in the quantitative agreement with this model if we assume that K_C and k_d have correspondent values listed in Table 10.1.

TABLE 10.1 Kinetic Parameters of the 1,1,3-trimethyl-3-(4-methylphenyl)butyl hydroperoxide decomposition activated by Et_4NBr

T, K	$k_d \times 10^5$, s^{-1}	K_C, dm^3 mol^{-1}	E_a, kJ mol^{-1}	$\Delta_{comp}H$, kJ mol^{-1}
373	2.35 ± 0.07	34 ± 2		
383	5.25 ± 0.09	27 ±3	88 ± 5	−19 ± 2
393	10.0 ± 0.6	24 ± 2		

It should be noted that k_d values do not depend on the ROOH and Et_4NBr concentration and correspond to the *ultimate case when all hydro peroxide molecules are complex-bonded and further addition of* Et_4NBr to the reaction mixture will not lead to the increase of the reaction rate. *Values of equilibrium constants of* complex formation (K_C) between ROOH and Et_4NBr estimated are within 24–34 dm^3 mol^{-1} and typical for the hydro peroxides quaternary ammonium salts systems [1, 2, 4]. Estimated value of complex formation reaction enthalpies (ΔH_{com} in Table 10.1) corresponds to the hydrogen bond energy in weak interactions [10].

It should be noted that within proposed scheme determined kinetic parameters are effective values, since the model used does not consider the possibility of the formation of cation-hydroperoxide as well as anion-hydroperoxide complexes in the system and their simultaneous decomposition.

Investigation of the 1,1,3-trimethyl-3-(4-methyl) phenyl butyl hydro peroxide interactions with tetra ethyl ammonium bromide by NMR[1]H spectroscopy.

The interaction between 1,1,3-trimethyl-3-(4-methylphenyl) butyl and hydroperoxide tetra ethyl ammonium bromide can be observed by relative change of the chemical shifts in the NMR spectra. Thus the effect of Alk_4NBr on the signal position of hydro peroxide in the proton magnetic resonance spectrum has been investigated to confirm the complex formation between 1,1,3-trimethyl-3-(4-methylphenyl)butyl hydroperoxide and tetra ethyl ammonium bromide at 297–313K. There is no hydroperoxide decomposition in experimental conditions (rate constant of the hydroperoxide thermolysis reaction in acetonitrile is $k_{term}^{313} = 1.28 \times 10^{-10} c^{-1}$ [9]). In the NMR^{1}H spectrum of 4-CH$_3$-C$_6$H$_4$-C (CH$_3$)$_2$-CH$_2$-(CH$_3$)$_2$C-O-O-OH chemical shift at 8.49ppm is assigned to proton of the hydro peroxide group (-CO-OH). Changing of hydro peroxide concentration within 0.01–0.05 mol dm^{-3} does not lead to a shift of the hydro peroxide group proton signal in spectrum. Addition of an equivalent amount of tetra ethyl ammonium bromide to the system leads to a shift CO-OH group proton signal to the side of the weak fields without splitting or significant broadening.

Subsequent NMR^{1}H spectroscopic studies were carried out in conditions of the quaternary ammonium salt excess at 297–313K. The concentration of ROOH in all experiments was constant (0.02 mol·dm^{-3}), while the concentration of Et$_4$NBr was varied within the range of 0.1–0.3 mol·dm^{-3}. The monotonous shifting of the NMR signal with increasing of Et$_4$NBr concentration without splitting and significant broadening shows fast exchange between the free and bonded forms of the hydro peroxide. Such character of signal changing of the hydro peroxide group proton in the presence of Et$_4$NBr (Fig. 10.2) indicates the formation of a complex between hydro peroxide and Et$_4$NBr in the system. Thus, observed chemical shift of the -CO-OH group proton (d, ppm) in the spectrum of 4-CH$_3$-C$_6$H$_4$-C (CH$_3$)$_2$-CH$_2$-(CH$_3$)$_2$C-O-OH – Et$_4$NBr mixture is averaged signal of the free $(\delta_{ROOH}$, ppm) and complex-bonded $(\delta_{comp}$, ppm) hydro peroxide molecule.

A nonlinear dependence of the changes of the proton chemical shift Δd $(\Delta d = d - \delta_{ROOH})$ of hydro peroxide group on the Et$_4$NBr initial concentration (Fig. 10.2) was obtained. In conditions of Et$_4$NBr excess and formation of the 1:1 complex for the analysis of the experimentally obtained dependence the Foster – Fyfe equation can be used [11]:

$$\Delta d/[Et_4NBr] = -K_C\Delta d + K_C\Delta\delta_{max}, \qquad (4)$$

where K_C – the equilibrium constant of the complex formation between hydro peroxide and Et$_4$NBr, dm^3·mol^{-1}; $\Delta\delta_{max}$ – the difference between the chemical shift of the -CO-OH group proton of complex-bonded and free hydro peroxide $(\Delta\delta_{max} = \delta_{comp} - \delta_{ROOH})$, ppm.

This dependence of the Δd on the Et$_4$NBr initial concentration is linear (Fig. 10.2) in the coordinates of Eq. (4). The equilibrium constants of the complex formation between hydro peroxide and Et$_4$NBr (K_C) and the chemical shift of the -CO-OH group proton of complex-bonded hydro peroxide were determined and listed in Table 10.2.

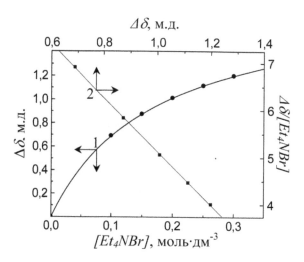

FIGURE 10.2   Dependence of change of the proton chemical shift of the 1,1,3-trimethyl-3-(4-methylphenyl)butyl hydroperoxide on the Et$_4$NBr initial concentration in direct (*1*) and Foster-Fyfe Eq. (*2*) coordinates ([ROOH]$_0$ = 0.02 mol·dm^{-3}, [Et$_4$NBr] = 0.1–0.3 mol·dm^{-3}, 297 K, CD$_3$CN).

TABLE 10.2   The Thermodynamic Parameters of Complex Formation Between the 1,1,3-trimethyl-3-(4-methylphenyl) Butyl Hydro Peroxide and Et$_4$NBr

T, K	K_c, dm^3·mol^{-1}	δ_{comp}, ppm	$\Delta_{comp}H$, kJ mol^{-1}	$\Delta_{comp}S$, J·mol^{-1}·K^{-1}	$\Delta_{comp}^{297}G$, kJ mol^{-1}
297	5.70 ± 0.07	10.35 ± 0.99			
303	4.74 ± 0.41	10.52 ± 1.00	−17 ± 3	−43 ± 10	−4.23
308	4.15 ± 0.18	10.50 ± 0.92			
313	4.06 ± 0.09	10.49 ± 0.92			

The chemical shift of the complex-bonded hydroperoxide δ_{comp} weakly dependent on temperature and is in the range 10.33–10.52ppm (Table 10.2). Enthalpy of the complex formation hydroperoxide-Et$_4$NBr is 17kJ mol^{-1}, which is consistent with the kinetic data (Table 10.1). The magnitude of the Gibbs free energy of complexation is negative, indicating the complex stability in this system.

Molecular modeling of the 1,1,3-trimethyl-3-(4-methylphenyl) butyl hydro peroxide decompositions activated by the Et$_4$NBr.

On the base of experimental facts mentioned above we consider the possible structure of the reactive hydro peroxide–catalyst complex. Model of the substrate separated ion pair (SubSIP) is one of the possible realization of join action of the salt anion and cation in hydro peroxide molecule activation. In this complex hydro peroxide molecule is located between corresponded cation and anion species. The solvent effect can be considered by means of direct inclusion of the solvent (acetonitrile) molecule to the complex structure. From the other hand methods of modern computer chemistry allow estimation the solvent effect in continuum solvation models approximations [8, 12].

Catalytic activity of the Et$_4$NBr in the reaction of the hydro peroxide decomposition can be considered due to chemical activation of the hydro peroxide molecule in the salt presence. Activation of a molecule is the modification of its electronic and nucleus structure that leads to the increase of the molecule reactivity. The SubSIP structural model has been obtained for the complex between hydro peroxide molecule and Et$_4$NBr (Fig. 10.3). An alternative structural model of the hydro peroxide-Et$_4$NBr complexes with combined action of the cation, anion and the solvent molecule has been proposed (Fig. 10.3). In this case a hydro peroxide molecule interacts with solvent separated ion pair of the salt. Only salt anion directly attacks hydro peroxide moiety whereas cation regulates the anion reactivity. This model is in agreement with key role of the salt anion revealed in experimental kinetic investigations [1].

Investigation of the proposed structural models properties for considered aralkyl hydro peroxides has revealed that complex formation was accompanied by following structural effects: peroxide bond elongation on 0.02 Å as compared with nonbonded hydro peroxide molecules; considerable conformation changes of the hydro peroxide fragment; O-H bond elongation on 0.07 Å in complex as compared with nonbonded hydro peroxide molecules; rearrangement of electron density on the hydro peroxide group atoms.

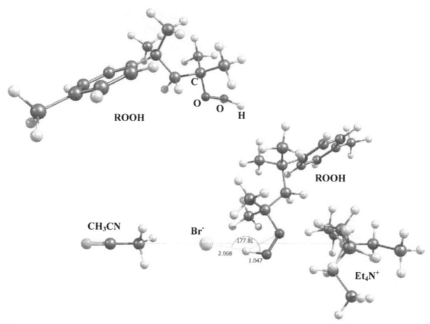

FIGURE 10.3   3D structural models of the 1,1,3-trimethyl-3-(4-methylphenyl) butyl hydro peroxide and the hydro peroxide-Et$_4$NBr complex with combined action of the cation, anion and the solvent molecule obtained by AM1/COSMO method.

Optimization of complex bonded hydro peroxides in COSMO approximation has shown that solvent did not affect on the character of structural changes in hydro peroxides molecules in the complexes. Only partial electron density transfer from bromide anion on to peroxide bond is less noticeable in this case.

Hydro peroxide moiety takes part in the formation of hydrogen bond (O) H...Br$^-$ in all obtained complexes. In all structures distances (O)H...Br$^-$ are within 2.03 , 2.16 Å, bond angle O-H...Br$^-$ is higher than 90° and within 170° , 180° corresponding from aralkyl moiety in the hydro peroxide molecule. Thus the interaction type of bromide anion with hydro peroxide can be considered as hydrogen bond [10].

Associative interactions of a hydro peroxide molecule with Et$_4$NBr to the peroxide bond dissociation energy (D$_{O-O}$) decrease. D$_{O-O}$ value for the aralkyl hydro peroxide was calculated according to equation (5) and for the ROOH – Et$_4$NBr SubSIP complex according to equation (6).

$$D_{O-O} = (\Delta_f H^0(RO^x) + \Delta_f H^0(^xOH)) - \Delta_f H^0(ROOH), \qquad (5)$$

$$D_{O-O} = (\Delta_f H^0(RO^x) + \Delta_f H^0(^xOH)) - \Delta_f H^0(ROOH_{comp}), \qquad (6)$$

where $\Delta_f H^0(RO^x)$ – the standard heat of formation of the corresponding oxi-radical; $\Delta_f H^0(^xOH)$ – the standard heat of formation of the ^xOH radical; $\Delta_f H^0(ROOH)$ – the standard heat of formation of the corresponding hydro peroxide molecule; $\Delta_f H^0(ROOH_{comp})$ – the standard heat of formation of the hydro peroxide molecule that corresponds to the complex configuration. D_{O-O} value for the hydro peroxide configuration that corresponds to complex one is less than D_{O-O} for the nonbonded hydro peroxide molecule. Difference between bonded and nonbonded hydro peroxide D_{O-O} value is $\Delta D_{O-O} = 44 kJ\ mol^{-1}$ and in accordance with experimental activation barrier decreasing in the presence of Et_4NBr: $\Delta E_a = 36 kJ\ mol^{-1}$.

Complex formation with structure of substrate separated ion pair is the exothermal process. Part of the revealed energy can be spent on structural reorganization of the hydro peroxide molecule (structural changes in –COOH group configuration). It leads to the corresponding electron reorganization of the reaction center (peroxide bond). Thus the increase of the hydro peroxide reactivity occurs after complex-bonding of the hydro peroxide molecule. So the chemical activation of the hydro peroxide molecule is observed as a result of the hydro peroxide interaction with Et_4NBr. This activation promotes radical decomposition reaction to proceed in mild conditions.

In the framework of proposed structural model the hydro peroxide molecule is directly bonded in complex with ammonium salt anion and cation. This fact is in accordance with experimental observations. Solvated anion approximation allows to directly accounting the solvent effect on the reactivity of complex bonded hydro peroxide.

10.4 CONCLUSIONS

Thus, a complex formation between 1,1,3-trimethyl-3-(4-methylphenyl)butyl hydroperoxide and Et_4NBr has been shown by kinetic and ^1H NMR spectroscopy investigations. Complexation enthalpy defined by these methods coincide. Complex formation enhances hydroperoxide reactivity. Lowering of the activation barrier for the complex-bonded hydroperoxide decomposition as compared with its thermolysis in acetonitrile is $36 kJ\ mol^{-1}$. A structural model of the hydroperoxide-Et_4NBr complex has been proposed. Formation of the complex with proposed structural features is accompanied with chemical activation of the aralkyl hydro peroxide molecule.

KEYWORDS

- Catalysis
- Complexation
- Hydro peroxide
- 1H NMR spectroscopy
- Kinetics
- Molecular modeling
- Tetraethyl ammonium bromide

REFERENCES

1. Turovskij, N. A., Antonovsky, V. L., Opeida, I. A., Nikolayevskyj, A. M., & Shufletuk, V. N. (2001). Effect of the Onium Salts on the Cumene Hydroperoxide Decomposition Kinetics, *Russian Journal of Physical Chemistry B.*, *20*, 41.
2. Turovskij, N. A., Raksha, E. V., Gevus, O. I., Opeida, I. A., & Zaikov, G. E. (2009). Activation of 1-Hydroxycyclohexyl Hydroperoxide Decomposition in the Presence of Alk$_4$NBr, *Oxidation Communications*, *32*, 69.
3. Matienko, L. I., Mosolova, L. A., & Zaikov, G. E. (2009). Selective Catalytic Oxidation of Hydrocarbons, New prospects, *Russ Chem Rev*, *78*, 221.
4. Turovskyj, M. A., Opeida, I. O., Turovska, O. M., Raksha, O. V., Kuznetsova, N. O., & Zaikov, G. E. (2006). Kinetics of Radical Chain Cumene Oxidation Initiated by α-Oxycyclohexylperoxides in the Presence of Et4NBr, *Oxidation Communications*, *29*, 249.
5. Hock, H., & Lang, S. (1944). Autoxidation of Hydrocarbons (VIII) Octahydroanthracene Peroxide, *Chem. Ber*, *77*, 257.
6. Vaisberger, A., Proskauer, E., Ruddik, J., & Tups, E. (1958). Organic Solvents. [Russian Translation] Moscow Izd Inostr. Lit.
7. Stewart, J. J. P., (2009). Mopac Stewart Computational Chemistry, Colorado Springs, CO, USA, http://OpenMOPAC.net.
8. Klamt, Cosmo, A. (1993). A New Approach to Dielectric Screening in Solvents with Explicit Expressions for the Screening Energy and its Gradient, *J chem. Soc Perkin Trans*, *2*, 799.
9. Turovsky, M. A., Raksha, O. V., Opeida, I. O., Turovska, O. M. (2007). Molecular Modeling of Aralkyl Hydro peroxides Homolysis, *Oxidation Communications*, *30*, 504.
10. Williams, D. H., & Westwell, M. S. (1998). Aspects of Week Interactions, *Chem Soc Rev*, *28*, 57.
11. Fielding, L. (2000). Determination of Association Constants (K_a) from Solution NMR Data, *Tetrahedron*, *56*, 6151.
12. Cossi, M., Scalmani, G., Rega, N., & Barone, V. (2002). New Developments in the Polarizable Continuum Model for Quantum Mechanical and Classical Calculations on Molecules in Solution, *J. Chem. Phys.*, *117*, 43.

CHAPTER 11

PROMISES OF PERSONALIZED MEDICINE IN 21ST CENTURY

SANJAY KUMAR BHARTI and DEBARSHI KAR MAHAPATRA

CONTENTS

ABSTRACT

Every individual is unique and uniqueness includes clinical, genomic and environmental information, as well as the nature of diseases, their onset, duration and response to drug treatment. Personalized medicine is an individualized therapy which takes into account the unique profiles of a single patient for optimized treatment. The discovery of genetic, genomic and clinical biomarkers have revolutionized the treatment option in the form of personalized medicine (PM) which allows us to accurately predict a person's susceptibility/ progression of disease, the patient's response to therapy, and maximize the therapeutic outcome in terms of low/no toxicity for a particular patient. It aims to provide right treatment, to the right patient, at the right time, at right cost. "One size does not fit all" is the prime reason for the evolution of PM, which emphasizes on genetic makeup of individuals that can be correlated with difference in drug therapy. One of the classic example include: "same symptoms, same findings, and same disease → different patient → same drug, same Dose → different effects; either therapeutic or subtherapeutic or toxic." This reflects the need of PM for better health outcome. Information derived from genetic tests using biomarkers are used for rapid disease diagnosis, risk assessment and better clinical decision. It plays an important role in reducing/ avoiding the adverse drug reactions and optimizing drug dose by identifying drug responders and nonresponders. The review describes the human genetic variations, biomarkers, challenges, regulation, barriers and impact of personalized medicine on clinical practice.

11.1 INTRODUCTION

The current approach of treatment considering "one-size-fits-all" may be one of the major causes of treatment failure [1]. The clinical decisions are mainly taken on the basis of the clinical stage(s) of the disease, the patient's medical and family history, and cost of treatment but ignoring the individual's biological or genetic make-up. While these approaches did achieve remarkable therapeutic efficiency in curing certain cancers (e.g., testicular cancer and pediatric leukemia), the overall survival rate of cancer patients has improved little in the last decades, arguing for the need of novel ideas and strategies in disease management. The new treatment decision should integrate comprehensive path physiology of disease, right therapeutic approach and drug delivery vehicle and other personal attributes (age, sex, nutrition, psychological conditions and gut micro biome). Such integration has brought the concept of personalized medicine that addresses the right drug to the right person at the right time.

Personalized medicine emphasizes individualized treatment by considering the uniqueness of a person, a disease and a drug at specific time points where by delivering individualized health care to patients to get maximum therapeutic efficiency at the cost of the minimal adverse effects. Personalized medicine has been widely and successfully applied to the management of numerous human diseases such as cancers and diabetes etc. The best example of personalized cancer therapy comes from the identification and clinical testing of the estrogen receptor (ER) in breast cancer and BCR/ABL in CML. Despite the impressive clinical benefits from this individualized health care, the current chemotherapies still have dismal outcomes. In such cases, further disease characterization using suitable biomarkers (e.g., genetic and epigenetic alterations) may aid the clinicians to decide the best means to deliver the drug in order to provide the maximal efficiency. Among many biomarkers, epigenetic modifiers are the most essential players in the design of personalized medicine, because these markers are tightly correlated to individualized attributes.

A Personalized Medicine (PM) is an approach to provide better health services that involves integrating genomics technologies and advances with clinical and family histories in order to provide more coherently tailor therapeutics to individual patients. A key component of personalized medicine is translating the science of pharmacogenomics into clinical practice [2]. Its purpose is to deliver the right treatment, to the right patient, at the right time every time. It is giving the opportunity to bring healthcare to a new level of effectiveness and safety. Personalized medicine involves the use of newer techniques of molecular analysis to manage the disease in efficient manner or to predispose toward a disease. It aims to achieve optimal medical outcomes by using patient's unique genctic and environmental profile and thus helping physicians and patients to choose most suitable approaches for management/prevention of diseases/disorders. Such approaches may include genetic screening programs, which more precisely diagnose diseases and their subtypes, and thereby helping the physicians to select the most appropriate type and dose of medication best suited to a certain group of patients [3].

11.2   BACKGROUND OF PERSONALIZED MEDICINE

Every individual is unique and uniqueness includes clinical, genomic and environmental information, as well as the nature of diseases, their onset, duration and response to drug treatment. Such uniqueness should be used to treat/ manage the disease effectively. Personalized medicine is an individualized or individual-based-therapy, which takes into account the unique profiles of a single patient for optimized treatment. The discovery of genetic, genomic and clinical biomarkers have revolutionized the treatment option in the form of personalized medicine which allows us to accurately predict a person's susceptibility to disease, the progression of disease, the patient's response to therapy, and maximize the therapeutic outcome in terms of low toxicity or no toxicity for a particular patient. Personalized medicine also requires the integration of other physiological factors, such as age, race, sex, life style, diet, genetic background, drug metabolism and health condition into therapeutic decisions [4]. Utilization of such specific marker based therapy can avoid unnecessary medication-associated toxicity and costs. The rate of drug metabolism influences the clearance of drug from systemic circulation; consequently dosage regimen should be decided accordingly in patient. Even with the similarity in drug metabolism rate, gut micro biome in specific patient can significantly affect the therapeutic efficiency [5]. In cancer patients, individual biomarkers became important to allow physicians to take clinical decisions and deliver more personalized care to a specific patient (Fig. 11.1).

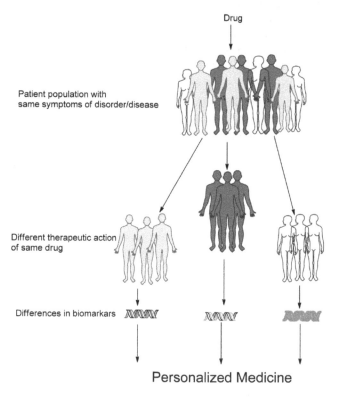

FIGURE 11.1   Application of Personalized Medicine.

Genome studies have revealed the presence of specific molecular bio-markers in the body, which are relative to specific gene variations. They usually confirm the clinical diagnosis in the presence of a disease. They may also indicate that a person in a dormant state is susceptibility to a specific disease and/or more reliably predict a person's differential response to treatment. Such biomarkers are indicative as the basis of new genomics-based predictive and diagnostic tests and the results of which may be used by the physicians to:

- firstly, diagnosis of the particular disease occurred/present in an individual;
- secondly, find individual's risk to the particular disease from which s/he is suffering;
- thirdly, identify whether a particular line of therapy will prove best for the particular patient or not, or finding the best alternative; and
- fourthly, tailor dosing regimens to individual variations in metabolic response.

The success of PM depends upon:

- Which biomarkers are to be selected for diagnosis?
- Which analytical assay will serve the purpose?
- Will both above stated points hold good-if yes, positive treatment with targeted drugs is possible.

11.2.1 HUMAN GENETIC VARIATIONS

In last decade, our understanding of human disorders/diseases has advanced significantly with sequencing of the genomes of humans and other species. Technological advancement in analyzing millions of genes and inter genetic variants in the form of single nucleotide polymorphisms (SNPs) and copy number variants (CNV) per individual have accelerated our understanding of individual differences in genetic make-up. Genome-wide association studies (GWAS) have successfully identified common genetic variations associated with numerous complex diseases. The contribution of environment gene-environment interactions, gene-drug relationship and epigenetic has been shown in therapeutic outcome. The sequencing technologies are advancing rapidly and now it is possible to identify rare variants in the patient but the cost of individual whole genome sequencing is still high. The application of genomics in personalized medicine includes its potential to benefit/risk assessment, diagnosis, prognosis and treatment and/or management. Some examples like BRCA1/BRCA2 testing for risk assessment of breast cancer, gene expression profiles to diagnose breast cancer subtypes, prediction of the severity of Fragile X syndrome by number of tri nucleotide repeats and response of trastuzumab (herceptin) to HER2-positive women in breast cancer and CYP2C19 variants are associated with diminished clopidogrel response [6].

11.2.2 GENETICS AND HUMAN DISEASES

Genetic disorders are classified into several major groups. The first group comprises chromosomal disorders such as Down syndrome, which is caused by an extra copy of chromosome 21. The second group consists of single gene disorders, such as cystic fibrosis and sickle cell anemia. The majority of genetic diseases, however, are multi factorial in nature. Indeed, rather than being associated with changes in only one or a few genes or proteins, many diseases are likely a manifestation of multiple interconnected aberrant pathways and numerous molecular abnormalities. Defects such as cleft lip and neural tube defects, heart disease, diabetes, and cancer result from a combination of multiple genetic and environmental causes [7]. The linkage analysis, candidate gene association studies, and GWASs are the methods currently used to identify phenotypic features of diseases and disease-susceptibility loci. Linkage analysis is useful for identifying familial genetic variants that have large effects and has been successfully used to discover several mutations responsible for monogenic forms of disease, such as maturity-onset diabetes of the young (MODY) heterozygous mutations in the Glucokinase (GCK) gene were shown to cause MODY2 [8], whereas mutations in the hepatocyte nuclear factor-1β (HNF-1β) gene were shown to be related to the development of MODY5 [9]. Genes associated with disease can also be identified on the basis of association testing in populations rather than in families. The methods include candidate gene association studies and GWASs. Candidate gene association is based on measurements of selected biomarkers from relevant path physiological pathways. The scores of candidate genes related to T2DM have been investigated using this approach and the PPARG and KCNJ11 genes were found to be directly linked to the development of the disease. The PPARG gene encodes the peroxisome proliferator-activated receptor-γ, a type II nuclear receptor that plays an important role in adipogenesis and insulin sensitivity by regulating the transcriptional activity of various genes. The KCNJ11 gene, located on the short arm of chromosome 11, encodes the pore-forming subunit of the ATP-sensitive potassium channel Kir6.2 in pancreatic β cells. Gain-of-function mutations in KCNJ11 open the potassium channel and inhibit the depolarization of β cells, leading to a defect in insulin secretion.

In addition, linkage analysis has demonstrated that the presence of the TCF7L2 gene increases the risk of developing T2DM in almost all ethnic groups. However, risk allele frequencies of SNPs in TCF7L2 in European populations were shown to be higher (40%) than those in Japanese (5%), indicating that TCF7L2 variants have little effect on T2DM susceptibility in the Japanese population.

In recent years, researchers have discovered numerous genes with variations in sequence or expressions that contribute to disease susceptibility and some of them provide the basis for molecular targeting of some diseases. The results from other studies have also indicated that genetic variability may be responsible for heterogeneous patient responses to treatment.

11.2.3 GENETIC TESTING

The genetic information of an individual can provide clues for progressive treatment in patient often obtained through genetic testing which primarily involves certain measurable molecular indicators known as "biomarkers." Genetic testing are of two types: somatic and germ cell genetic testing, based on the type of DNA mutations involved and their effect on disease cause. Somatic cell genetics refers to mutations in the DNA of somatic cells some time after conception and is therefore not heritable. Germ cell genetics refers to mutations in the DNA of the germ cells (ova or sperm) and so are heritable.

11.3 BIOMARKERS

Biomarkers are measurable molecular indicators in an individual's body, which describes current health status or predictors of susceptibility to disease of an individual and provide evidences for effectiveness of treatment. Genetic information is one of many such indicators. Other biomarkers include the products of genes or metabolic activity, such as proteins, hormones, RNA or other signaling molecules used by cells. Mainly biomarkers are used in clinical practice to:
- identify the probable risk of disease;
- diagnosis and accessibility of severity of existing disease;
- stratify patients to potentially tailor treatment;
- understanding molecular mechanisms of disease;
- improved postmarket surveillance;
- a clue for drug discovery.

11.3.1   BIOMARKERS FOR PERSONALIZED MEDICINE

Biomarkers of diseased organ are indicators of disorders, which are mostly determined by the interaction or coordination of diverse signaling pathways and a result of the combined tumoral, stromal and inflammatory factors of the original heterogeneous tumor. The molecular biomarkers may be mutant genes, RNAs, proteins, lipids, carbohydrates and metabolites, whose expression levels are positively or negatively correlated to biological behaviors or clinical outcomes. Clinical biomarkers include diagnostic biomarkers indicating the presence or absence of a disease, and prognostic and therapy optimization biomarkers indicating patient and disease characteristics that influence treatment plans.

11.3.1.1   CHROMOSOMAL ABNORMALITIES

Chromosomal abnormalities may arise due to fragmentation and/or relocation of fragments of chromosome(s) and to other chromosomes. It can cause the loss or alteration of genetic functioning. The identification of patient-specific translocations has revolutionized the personalized cancer therapy. Various types of chromosomal abnormalities have also been identified in leukemia patients.

11.3.1.2   GENE MUTATIONS

Oncogenes and tumor suppressor genes (TSGs) are two main types of genes involved in cancer pathogenesis. Although, numerous proto-on co genes are very crucial to normal cell growth, when mutated become hyper-activate leading to cancer pathogenesis. Mutation in TSG has been evaluated in lung, colorectal, breast cancer and leukemia. The abnormalities of TP53 gene have been seen in more than half of human malignancies. The understanding of disease-gene relationship in individuals can be helpful in predicting the response to certain medications, thereby prompting individualized therapy. Mutations of some well-known oncogenes include DNA methyl transferases (DNMT3a, DNMT1), protein kinases (e.g., EGFR, KIT and FLT3, ErbB2) and GTPase (e.g., KRAS) are some examples of gene mutation. Designing of inhibitors keeping in mind the mutation occurred in particular gene will be more effective strategies. The identification of KIT and FLT3 mutations and designing of their specific inhibitors for specific patients have advanced the personalized cancer management. In clinics, specific KIT mutation in codon 822 can predict the response to imatinib, whereas mutation in codon 816 is not and can be treated successfully with midostaur in or dasatinib. Specific mutations in the EGFR gene as predictors of response to erlotinib or gefitinob are the first example of using biomarkers to predict response to targeted agents. Discovery of chemotherapy-induced gene mutations can also help to develop specific inhibitors and also to predict how the individual patients will response to the selected drugs.

11.3.2   SINGLE NUCLEOTIDE POLYMORPHISMS (SNPS)

Single nucleotide variations occur at a specific site in coding and noncoding regions of the genomic DNA. SNPs are taking place every 100 to 300 base pairs along the human genome and making up about 90% of all human genetic variation. In the past two decades over 10 million SNPs have been discovered. While most SNPs have no obvious impacts on critical biological processes, some of them did implicate disease phenotypes or affect patients' response to a drug therapy thereby representing favorable biomarkers for diagnosis and prognosis. Iri note can is a chemotherapeutic agent used for colon carcinoma patients. Single nucleotide polymorphism affects the iri note can metabolism, which can predict whether the specific patient subgroup responses to iri note can therapy. SNPs in BRCA1 and TP53 genes are associated with increased risk for breast cancer, whereas other SNPs, like rs2352028 at chromosome 13q31.3, indicate higher susceptibility for lung cancer in never-smokers. Polymorphism in cytochrome P450 (CYP) genes, including CNV, mutation, insertion and deletion have significantly affected individual drug response. The utilization of CYP polymorphism biomarkers in clinics can improve drug response and diminish treatment costs.

Recently SNPs have been identified in mircoRNA (miR) genes which function as another indicator for cancer risk and therapeutic outcome of drug in individual patients. Abnormal miR expression causes various human diseases and may be used as biomarker to classify cancer types. SNPs of miRs could significantly modulate their biological activities thereby representing attractive biomarkers for making clinical decisions. Overall, it is increasingly clear that SNPs will be an important biomarker to design and develop personalized medicine. Due to clinical potential of SNPs, several groups, like the U.S. Human Genome Project (HGP) and SNP Consortium are creating human genome SNP maps in order to identify useful prognostic and diagnostic biomarkers. However, the largest barrier for SNPs' clinical applications in individualized medicine is the genetic discrimination issue.

11.3.2.1   CIRCULATING DNA, RNA AND TUMOR CELLS

Various types of circulating RNA and DNA in either free or bonded form are important biomarkers. DNA can be used as an early marker for cancer detection due to its enhanced level and for therapeutic monitoring. miRs can activate or suppress target gene expression and are the largest class of gene regulators. In circulating blood, differential expression of miRs between normal and diseased people has been evidenced. For example, the level of miR-155, miR-210 and miR-21 in B-cell lymphoma patients' serum is elevated. Mitchell et al. reported that serum levels of miR-141 can distinguish patients with prostate cancer from healthy controls. These circulating miRs are emerging as important noninvasive biomarkers which can benefit the diagnosis, prognosis and therapeutic efficacy. Additionally, circulating cancer cells can provide clinicians with a more advanced noninvasive way to know the stage of cancer metastasis, the individualized approaches for patients, and the way to monitor disease progress. In general, DNA, RNA, miR and circulating cancer cells are distinct between cancer and normal cells, even the cancer cells at different stages of disease progression. Such diversified biomarkers offer a crucial platform to understand the etiology and pathology of cancers with a strong support to personalized medicine.

11.3.2.2   "OMICS" FOR THE DESIGN OF PERSONALIZED MEDICINE

"Omics," including genomics, transcriptomics, proteomics, metabolomics and epigenomics, data offer the most comprehensive information for the design of personalized medicine through the utilization of high throughput technologies. It can discover and characterize tumor cells, tumors and person-specific biomarkers that can help in better individualized therapy. However, the application of omics data in personalized cancer therapy is very challenging. Therefore, very few have been translated from bench to bedside. This is why, instead of using omics data in making treatment decision, currently physicians prefer to adopt simple biomarkers, such as kinase mutation, IHC staining, hormone receptor expression and/or chromosomal aberrations.

11.3.3   PHARMACOGENOMICS IN PERSONALIZED MEDICINE

Genome-guided therapy taking the advantage of individual patients' genomic markers to ascertain whether a person is likely to respond to a given therapy, avoid toxic side-effects, and adjust the dosage of medications to optimize their efficacy and safety. Genetic variations in humans are recognized as an important determinant of drug response variability. Different patients respond differently to the same drug, with genetics accounting for 20–95% of the variability. Pharmacogenomics is the study of how human genetic variations affect an individual's response to drugs, considering the drug's pharmacokinetic properties like absorption, distribution and metabolism. It plays an important role in reducing/avoiding the adverse drug reactions (ADRs) and optimizing drug dose by identifying drug responders and nonresponders. Recently, the U.S. Food and Drug Administration (FDA) has realized the contribution of pharmacogenomic in better healthcare and advocated the consideration of pharmacogenomic principles in making safer and more effective drug. In order to improve the quality of already marketed drugs, the FDA has updated certain drug labels to include pharmacogenomic information. Currently, more than one hundred FDA approved drugs have pharmacogenomic information on their labels that describe gene/allele responsible for drug response variability and risk for adverse events. One of the well-known examples of pharmacogenomic specific drug-gene relationships is warfarin-CYP2C9 [10]. Gene CYP2C9 encodes an important CYP enzyme that plays a major role in metabolism of more than 100 therapeutic drugs. The genetic polymorphisms of CYP2C9 are associated with altered enzyme activity leading to toxicity at normal therapeutic doses of warfarin. Understanding how the genetic variants contribute to various drug responses is an essential step of personalized medicine. The success of personalized drug treatment largely depends on the availability of accurate and comprehensive knowledge of pharmacogenomic-specific drug–gene relationships, such as warfarin-CYP2C9 and irinotecan-UGT1A. With technological advancement, the cost and duration of genomic sequencing continue to decrease sharply and there is intensive research aimed at understanding how the changes that occur within the genome can alter its function and the genomic variations that constitute individual susceptibility to diseases and responses to therapy. The connection of a personal genome with the personal medical record of patients has a potential to improve prediction and prevention and to allow a more proactive therapeutic strategy. It is evident that pharmacogenomic and individualized drug therapy are the building blocks of personalized medicine. A growing number of drugs are now used for the treatment of cancer in subjects selected by a companion genetic test. Personalized medicine while based upon genomic knowledge of the individual requires equally essential personalized environmental information as well as the understanding of every subject's capacity for health-promoting behavior.

One of the examples of genetic polymorphisms of drug targets is variants located in the enzyme vitamin K epoxide reductase complex subunit 1

(VKORC1), the target of warfarin the most commonly used oral anticoagulant worldwide [11]. Inhibition of VKORC1 by warfarin leads to vitamin K depletion and, consequently, the production of coagulation factors with reduced activities. Several mutations in the coding region of VKORC1 have been identified. Most of them are rare and have been associated with resistance to warfarin, necessitating larger dosages of warfarin. More common variants are located in the introns and regulatory regions of the VKORC1 gene and influence warfarin dosage within the normal range. Cytochrome P450 family members are important enzymes that metabolize most, if not all, clinical drugs. With CYP450 polymorphisms, subjects can metabolize the medication too rapidly (ultra-metabolizers), rendering it ineffective, or too slowly (poor metabolizers), leading to increased plasma drug concentration, potentially causing adverse reactions and, in contrast, in case of pro drugs, ineffective activation. Without knowing the personal characteristics of patients, their clinical and genetic information, it may take months of trial-and-error testing to find the right drug dosage for each of them. As many patients are either ultra-or poor-metabolizers, pharmacogenetic testing will help to identify these differential responders, providing a basis for personalized drug treatment of a large number of patients. For example, CYP2D6, one of the best-studied drug-metabolizing enzymes, acts on 25% of all prescription drugs. About 10% of the general population has a slow-acting form of this enzyme, and 7% have a super-fast acting form. In total, 35% are carriers of an abnormally functional 2D6 allele. This is important in cancer treatment as tamoxifen is one of the best-known anticancer drugs that CYP2D6 metabolizes [12]. A growing number of drugs have companion diagnostics, and about 10% of marketed medications propose or recommend genetic testing for optimal treatment. Drug toxicity can be influenced by genetic factors known to affect drug responses such as drug targeting, drug metabolizing-enzymes, drug transporters and genes, indirectly impacting drug action. "On-target" toxicity occurs with drug accumulation in the circulation, as in the case of bleeding with high warfarin doses, while "off-target" toxicity is observed, with stat in-induced myopathy. The mechanism of stat in-induced myopathy is still unclear and generally rare: 1/10,000 cases per year on standard therapy. However, the incidence increases with higher stat in doses or concomitant use of other drugs, such as cyclosporine or amiodarone [13]. Simvastatin is administered as an inactive lactone pro drug that is converted to the active metabolite simvastatin acid in plasma, liver and intestinal mucosa via nonenzymatic and carboxyl esterase-mediated processes. The organic anion transporting polypeptide1B1 encoded by SLCO1B1 is an influx transporter on the base lateral membrane of hepatocytes. It facilitates the uptake of statins and other drugs. A common

SNP of SLCO1B1 (rs4149056) has been shown to markedly affect individual variation of stat in pharmacokinetics in healthy volunteers. The plasma concentration of simvastatin acid was shown to be increased by 2- to 3-fold in carriers of the CC homozygous genotype versus the TC heterozygous or TT homozygous genotype. Strong association was detected with SNP located in one intron of the SLCO1B1 gene that gave an odd ratio of 3.4 per copy of the C allele and 17.4 among CC homozygous compared to TT homozygous genotypes. The study suggested that genotyping of SLCO1B1 may help in identifying persons with abnormal SLCO1B1 and thereby increase the safety and efficacy of statin therapy. These examples of successful pharmacogenetic testing, unraveling the basis of some individual drug responses by single-gene polymorphisms, can guide future research in pharmacogenomic and its application. Realistically, pharmacogenomic will require complex, polygenic and gene-environment considerations to prove its clinical utility. Research into gene pathways and networks, the integration of genetics and genomics, proteomics, metabolomics and epigenetics, noncoding RNA derived from hypothesis-free investigation in GWAS and prospective clinical trials evaluating the utility and cost-effectiveness of genetic testing in drug therapy are a few examples of exploration that should continue.

11.3.4   WARFARIN-CYP2C9 GENE

Warfarin, an anticoagulant drug, is widely used to prevent thrombosis, but hemorrhagic complications are seen in some patients. Two genes, Cytochrome P450–2C9 (CYP2C9) and vitamin K epoxide reductase (VKORC1), are associated with the pharmacokinetics and pharmacodynamics of warfarin. The metabolism of (S) warfarin to its inactive metabolite is mediated by CYP2C9. Persons with at least one copy of either CYP2C9*2 or CYP2C9*3 require less warfarin for effective anticoagulation activity. The hemorrhagic complications are more common in persons carrying the CYP2C9*2 or CYP2C9*3 allele. The function of VKORC1 is the gamma carboxylation of glut amyl residues, which is required for the activation of multiple vitamin K-dependent clotting factors. A functional variant in the promoter region of the VKORC1 gene was found to alter a transcription factor-binding site. The CYP2C9 and VKORC1 polymorphisms account for 18% and 30% of the observed variability of warfarin, respectively. Specifically, the VKORC1 variant predicts warfarin sensitivity, and the CYP2C9*2/*3 variants affect warfarin clearance. Clinical factors including age, sex, body size, race/ethnicity, smoking status, and relevant concomitant medications are responsible for an additional 12%

of dose variability. Overall, VKORC1 and CYP2C9*2/*3 variants as well as clinical factors account for up to 60% of inter individual variability.

11.3.5   HERCEPTIN TREATMENT OF BREAST CANCER AND THE HER2/NEU GENE

About 18–20% of patients with breast cancer show amplification of the HER2/neu gene or over expression of its protein product. These factors are associated with increased risk of disease recurrence and poor prognosis. The development of HER2-targeted therapies for patients with HER2-positive disease has dramatically reduced the risk of recurrence after the initial therapy and has led to improved prognosis [14]. Various HER2-targeted drugs are approved or in development, such as monoclonal antibodies (mAbs) that are directed against its external domain (e.g., trastuzumab and pertuzumab), small molecule tyrosine kinase inhibitors (e.g., lapatinib and neratinib), anti-HER2 antibodies conjugated to toxic molecules (e.g., trastuzumab-DM1 or TDM1), and chaperone antagonists (e.g., geldanamycin) [15]. Herceptin (trastuzumab) is a recombinant humanized IgG1-kappa monoclonal antibody that selectively binds with high affinity to the extracellular domain of HER2. Based on results from randomized clinical trials, trastuzumab-containing regimens are now recommended for women with HER2-positive metastatic breast Cancer [16]. Data from trials of first generation adjuvant regimens combining trastuzumab with various chemotherapeutic drugs showed significant improvements in disease-free and overall survival rates [17]. Studies on second-generation adjuvant regimens comprising other HER2-targeted agents such as lapatinib and pertuzumab are underway, and newer drugs such as T-DM1 and neratinib are being actively tested in the metastatic setting.

11.3.6   CARBAMAZE PINE THERAPY AND THE HLA-B*1502 ALLELE

Carbamaze pine is an important drug for the treatment of seizure, bipolar disorder, trigeminal neuralgia, and chronic pain. However, carbamazepine is also associated with hypersensitivity reactions that range from benign urticaria to life-threatening cutaneous disorders, including Stevense-Johnson syndrome (SJS) and toxic epidermal necrolysis (TEN). Both conditions are associated with significant morbidity and mortality. Recently, pharmacogenomic studies have found a strong association between carbamazepine-induced SJS and TEN and the HLA-B*1502 and HLA-B*5801 alleles in the Han Chinese populations in Taiwan [18, 19] and other Asian countries [20, 21]. Furthermore, the incidence of SJS and TEN among people treated with the drug is substantially reduced when individuals carrying HLA-B*1502 are excluded from carbamazepine therapy [22]. Patients of Han Chinese descent with molecular evidence that allele should, therefore, be treated with other classes of anti epileptic drugs. However, the allele frequency of HLA-B*1502 is markedly lower in Caucasians (1.2%) than in Han Chinese (8%). In addition, no association between HLA-B and carbamazepine-induced SJS and TEN has been found in Caucasian patients. Therefore, the United States Food and Drug Administration recommend genetic screening for patients of Asian ancestry before initiation of carbamazepine therapy.

11.3.7 ANTI DIABETIC DRUGS

The current drugs for diabetes include insulin, metformin, sulfonylurea, and thiazolidine diones. Metformin suppresses hepatic gluconeogenesis by activating AMP-activated protein kinas, which inhibits the expression of hepatic gluconeogenic genes PEPCK and Glc-6-pase by increasing the expression of the small hetero dimer partner. It was recently found that variations in the organic cation transporter 1 and 2 (OCT1 and OCT2), proteins responsible for the hepatic transport of metformin, might affect metformin response [23]. Sulfonylurea bind to ATP-sensitive potassium channels in pancreatic β cells, thereby stimulating insulin release in a glucose-independent manner [24]. Recently, polymorphisms in drug target genes (ABCC8, KCNJ11) and diabetes risk genes (TCF7L2 and IRS-1) have been shown to be associated with variability in the response to sulfonylurea drugs in patients with T2DM [25]. Pearson et al. [26] found that patients with hepatocyte nuclear factor-1α (HNF-1α) gene mutations were supersensitive to treatment with sulfonylurea but responded poorly to treatment with metformin. In addition, research has shown that individuals with the TCF7L2 risk genotype respond poorly to treatment with sulfonylurea [27]. Variations in CYP2C9, an enzyme involved in sulfonylurea drug metabolism, KCNJ11, Kþ inward rectifier Kir6.2, and the sulfonylurea receptors SUR1 and ABCC8 are all genetic modifiers of the response to treatment with sulfonylurea.

For thiazolidinediones, the focus has been on PPARγ variants, although the results are currently in conclusive. Despite these associations, the genetic data currently available are insufficient to support clinical decisions for the common forms of T2DM [28].

11.3.8   CYTOCHROME P450 ISO ENZYMES AND PERSONALIZED MEDICINE

Metabolism or biotransformation is the process whereby a substance is changed from one chemical to other by a chemical reaction in the body. For some drugs, the metabolite formed after transformation possesses therapeutic action. The process involves conversion of physical and chemical properties of drugs present in systemic circulation, which leads to generation of more polar moieties that are easily excreted by human body. There are many factors, which affects metabolic pattern among individuals and thereby arising the need of PM. Some of the variability are mentioned in the Tables 11.1–11.5.

TABLE 11.1   Gene/allele Responsible for Variation in Therapeutic Output of Drug(s)

Drug	Risk/Toxicity	Specific gene/allele
Diclofenac	Increased risk for hepatotoxicity	ABCC2 C-24T, UGT2B7×2
Atomoxetine, Thioridazine, Codeine, Tamoxifene	Toxicity	CYP2D6
Voriconazole	Hepatotoxicity	CYP2C19
Warfarin	Risk of bleeding	CYP2C9, VCORC1
Maraviroc	No response for CCR5-negative patients	CCR5
Imatinib	No response in cancer patients with absence of tumor activating c-Kit mutations	c-kit
Rasburicase	Hemolysis in G6PD-deficient patients	G6PD
Capecitabine	Orodigestive neutropenia	DPD
Erlotinib, Cetuximab, Panitumumab	No response in cancer patients with tumor EGFR-negative expression	EGFR
Anastozole Exemestane Letrozole Tamoxifene	No response in cancer patients with tumor ER-negative expression	ER
NSAIDs	Increased risk for cytolytic hepatitis	GSTM1/GSTT1

TABLE 11.1 *(Continued)*

Drug	Risk/Toxicity	Specific gene/allele
Tacrine	Increased risk for cytolytic hepatitis	GSTM1/GSTT1
Trastuzumab	No response in cancer patients with tumor HER2-negative expression	HER2
Carbamazepine	Severe cutaneous immunoallergic reaction, Increased risk for Stevens-Johnson syndrome (Asian)	HLA-B*1502
Abacavir	Increased risk for hypersensitivity reactions	HLA-B*5701
Nevirapine	Increased risk for hypersensitivity reactions	HLA Cw8-B14, HLA-DRB1×0101
Ximelagatran	Increased risk for cytolytic hepatitis	HLA-DRB1×0701
Flucloxacillin	Increased risk for cholestatichepatitis	HLA-DRB1*DQB1
Allopurinol	Increased risk for Stevens-Johnson syndrome	HLA-B*5801
Cetuximab, Panitumumab	No response in cancer patients with tumor specific K-RAS mutations	K-RAS
Isoniazid	Increased risk for cytolytic hepatitis	NAT2
Simvastatin	Increased risk for myopathy	SLC01B1×5
Flucloxacillin	Increased risk for cholestatic hepatitis	TNF-α −238G/A
Azathioprine6-Mercaptopurine, Thioguanine	Neutropenia	TPMT
Irinotecan	Diarrhea, Neutropenia	UGT1A1*28

TABLE 11.2 Enzymes Responsible for Variation of Drug Metabolism in Population, Necessitating the PM

Enzyme	Drug	Variability
Acetyl transferase	Isoniazid	Acetylation: slow or fast
Dihydropyrimidinedehydrogenase	5- Fluorouracil	Slow or inactivation
Aldehyde dehydrogenase	Ethanol	Metabolism: slow or fast
Catechol -O- methyl transferase	Levodopa	Methylation: slow or fast
Thiopurinemethyltransferase	Azathioprine	Methylation: slow or fast
Cholinesterase	Succinylcholine	Ester hydrolysis: slow or fast

TABLE 11.3 Variability in Therapeutic Effects Observed Due to Drug Metabolism in Individuals having Genetic Variability in CYP3A4 Isoforms

Drugs	Altered activity	Metabolism
Amiodarone	Anti-arrhythmic	N-deethylation
Carbamazepine	Anti-convulsant	Epoxidation
Codeine	Anti-tussive	N-demethylation
Cyclosporin	Immunosuppressant	N-demethylation
Dapsone	Anti-leprotic	N-oxide
Dextromethorphan	Anti-tussive	N-demethylation
Diazepam	Sedative	Hydroxylation
Diltiazem	Anti-arrhythmic	N-deethylation
Dolasterone	Anti-emetic	N-oxide
Erythromycin	Antibiotics	N-demethylation
Imipramine	Anti-depressants	N-demethylation
Isradipine	Anti-arrhythmic	Aromatization
Lidocaine	Anti-arrhythmic	N-deethylation
Lovastatin	Anti-lipidemic	Hydroxylation
Midazolam	Sedative	Hydroxylation
Nicardipine	Anti-arrhythmic	Aromatization
Quinidine	Anti-arrhythmic	Hydroxylation
Tamoxifen	Anti-neoplastic	N-demethylation
Theophylline	Anti-asthmatic	Oxidation
Tolterodine	Sedative	N-demethylation
Valproic acid	Anti-convulsant	Hydroxylation
Verapamil	Anti-arrhythmic	N-demethylation

TABLE 11.4   CYP450 Isoforms Influencing the Drug Action

Drug	Isoform
Carbamazepine	1A2, 3A4, 2C8, 2D6
Clotrimoxazole	3A4
Erythromycin	3A4
Griseofulvin	3A4
Isoniazid	2E1
Lansoprazole	3A4
Omeprazole	3A4
Phenobarbital	2B6, 2D6, 3A4
Phenytoin	2B6, 2C8, 2C19, 2D6
Rifampicin	2C8, 2C9, 2D6

TABLE 11.5   Drug Based on Biomarker for Better Cancer Therapy

Biomarker tests	Cancer form	Drug
BCR-ABL translocation	Chronic Myeloid Leukemia/ Acute Lymphoblastic Leukemia	Imatinib
KIT and PDGFRA mutations	GIST	Imatinib
HER2 amplification	Breast and gastric cancers	Trastuzumab, Lapatinib
KRAS mutations	Colorectal cancer	Panitumumab and cetuximab
EGFR mutations	Lung cancer	Gefitinib and erlotinib
EML4-ALK translocations	Lung cancer	Crizotinib
BRAF mutation V600E	Melanoma	Vemurafenib
ER	Breast Cancer	Tamoxifen, Letrozole
EGFR/*KRAS*	Colorectal Cancer	Cetuximab, Panitumumab
EGFR	Non-small cell lung Cancer	Gefitinib, Erlotinib
C-KIT (CD117)	Gastrointestinal stromal Tumor	Imatinib
TOP2A	Breast cancer	Epirubicin

11.4   MEASURES IN PERSONALIZED MEDICINE

11.4.1   OVERCOME DEFICIENCIES IN SCIENTIFIC KNOWLEDGE

Though PM is still in its infancy, scientific knowledge is growing very fast. It is therefore important that physicians should have adequate knowledge in the fields of epidemiology, medical genetics and medical statistics in order to make sense of the findings of PM. In addition, patients who seek information online expect their physicians to have an appropriate level of knowledge.

11.4.2   STRENGTHENING MEDICAL GENETICS

Genetics is the key components of PM. Physicians are increasingly expected to be able to engage in genetic analysis and counseling. Appropriate coverage and strengthening of medical genetics in basic medical education would therefore appear to be indispensable part of healthcare system.

11.4.3   DEVELOPING SUPER-SPECIALTY APPROACH

Because every disorder has its own distinct genetics and complexity, there is a need to promote appropriate specialist training and continuing education program, so that physicians have the opportunity to acquire PM-related knowledge in their own specialty. Patients, for their part, are entitled to know which physicians have the knowledge that is required to answer PM related questions.

11.4.4   PROVIDING MORE WEIGHT AGE TO THE FAMILY HISTORY

The importance of the family history should be promoted in medical education and training. There will be a need for clinical guidelines for its interpretation and usage, and efforts are to impart the necessary knowledge and skills to medical students.

11.4.5   PROMOTING DEVELOPMENTS AND CREATING TRANSPARENCY

Free access should be provided to PM-related services and they must be subjected to scientific review, so that potential users know what they should expect. This should include measures to ensure transparency regarding problems, conflicts of interest, faults of any kind. Developments, which cannot be regulated call for the adoption of clear positions on the part of the medical community.

11.4.6 IMPACT OF PERSONALIZED MEDICINE ON CLINICAL PRACTICE

Genetic testing and screening have already become common in certain specialties. The prenatal screening for some chromosomal abnormalities can now be done by examining fetal DNA circulating in the mother's blood. The adoption of personalized medicine as a standard of care in diagnosis, risk assessment and treatment has evolved rapidly.

In addition, it is important to highlight targeted therapeutics and companion diagnostics. In targeted therapeutics, usually drugs/biologics are designed to benefit a particular subpopulation. Companion diagnostics are accompanying laboratory tests and professional services identifying or measuring genes, proteins, or other substances that delineate the subpopulation that will derive benefit from the targeted therapeutic and yield important information on the proper course of treatment for a particular patient.

Genetic/genome testing services and targeted treatments are some of the fastest growing areas in health care. While new technology generally is associated with high cost, personalized medicine services have the potential to significantly reduce the cost and improve health outcomes by ensuring care is delivered earlier and with best treatment option that have a higher rate of success. It may also shorten the diagnostic processes for those with rare conditions or disease, thereby reducing costs. The capacity to sequence the whole genome of an individual at relatively low cost (possibly for $1000 by 2014), is a dramatic leap forward for the widespread clinical accessibility of these services. The low cost of sequencing coupled with our understanding of factors responsible for varied therapeutic efficacy indicates that the reach of personalized medicine into more areas of medicine has begun in earnest.

As a result of breakthroughs in the accuracy of whole genome sequencing as well as a dramatic reduction in the cost of sequencing, personalized medicine has graduated from testing services to clinically useful sequence analysis and whole genome sequencing. In particular, whole genome sequencing has become affordable and advances in interpretation and analysis have increased clinical uses.

11.4.7   FOCUSED THE RANOSTICS (DIAGNOSTICS AND THERAPEUTICS)

Development of new imaging techniques and genetic tools such as biomarkers and biochips.

11.4.8   GENETIC SCREENING OF PATIENTS

The impact on clinical decision making and health promotion in key areas including cancer, diabetes, obesity, heart disease, infectious disease and dementia-related conditions such as Alzheimer's disease. Advances in genomics and proteomics have led to more cost efficient discoveries, with physician more inclined to use targeted treatments.

11.5   CHALLENGES FOR PERSONALIZED MEDICINE

Scientific, regulatory, economic, educational, and policy barriers can hinder the implementation of PM. The value of personalized medicine may be increasingly obvious; the pace at which it will transform healthcare is not a certainty.

11.5.1   GENETIC PRIVACY AND NON-DISCRIMINATION

While genetic and genomic testing is expected to lead to better patient outcomes and improve medical care and health promotion, it could also lead to unintended and unwelcome consequences such as discrimination and unauthorized disclosures that harm patients and/or their families. In 2008, Genetic Information Non-discrimination Act (GINA) has been passed, which prohibits genetic discrimination in the health insurance market and in employment decisions. There is a pressing need to consider and expand the protections against genetic and genomic discrimination for individuals that undergo testing as well as a need for the adoption of a comprehensive framework that adequately addresses privacy concerns. Discrimination based on genetic information should be prevented.

Certain laws provide protection against misuse of genetic information (United States Federal Laws) but no laws have been amended in developing Nations. In many cases, public are prevented from full participation in personalized medicine related research or clinical care, unless full genetic privacy protections are ensured/maintained. Currently, laws are insufficient to protect against the misuse of genetic information. A number of US federal legislation provider are present for the protection of medical and genetic information in the United States, that would prohibit health insurers from discriminating against individuals on the basis of their genetic risk factors.

11.5.2 HEALTHCARE PROFESSIONALS' EDUCATION, AWARENESS AND ATTITUDE

Presently physicians are the decision-makers related to the application of personalized medicine in health care system. At present, developing Nations do not have the trained staffs to practice personalized medicine. Study syllabi for students have not incorporated personalized medicine concepts and tools as an important part. Physicians and other healthcare providers are the actual decision makers for introduction and application of genetic tests and pharmacogenomic based drugs. Establishing that the given therapeutic approach is the best suitable and the most appropriate cure for affected patient having genetic background and establishing biomarker based disease identification need to be further documented. This approach uses information systems for managing patient care, and deals with new ethical and legal issues that arise from molecular and genetic testing.

The large numbers of physicians are not proficient in genetics, genomics and molecular science, and lack confidence in dealing with genetic-related issues that arise in clinical settings. The number of board-certified physician geneticists and board-certified genetic counselors are less. The American Medical Association (AMA) council on science and public health report also noted that it is nearly impossible for most physicians to keep up to date on the increasing number of genetic/genomic tests. The former will only become more difficult as whole genome sequencing is leading rapidly to the discovery of gene associations that can be translated into diagnostic tools with clinical relevance.

The College of American Pathologists (CAP) has launched a determined effort to position pathologists as an essential leader of the medical team versed in genetic and genomic testing. However, the number of pathologists able to perform this function may not necessarily meet current need, which is already raising difficult challenges. While there are other professionals who may provide genetic counseling or test interpretation, physicians are ultimately responsible and accountable for the medical care a patient receives.

11.5.3   HEALTHCARE INFORMATION TECHNOLOGY (HIT)

Healthcare information technology is an online data retrieval system to link patient's genetic information for research purposes thereby enhancing the practice of new personalized therapies easier. There is a significant difference in actual implementation and proper use of HIT. System standardization, calibration, staff training, proper utilization and removal of IT investment barriers are required to ensure widespread adoption of HIT. It has a significant role in successful adoption of personalized medicine as standardized data structures are necessary to characterize and record diseases by their molecular profiles, allowing physicians to coordinate and optimize individual patient care based on molecular and genomic information. Integrated health system networks can lead to better-tailored treatments as responses are compared between patients with similar conditions and genetics, long-term health and economic outcomes are calculated, and biomarker-disease associations are identified and validated by linking scientific and clinical research. Decision-support software can provide an important tool, but the unchecked rise of such technology raises questions as well because the software may not contain the input or information by expert healthcare professionals. However, there is growing concern that as personalized medicine becomes mainstream medicine possibly within the decade, it will be dominated by individuals though rigorously trained, who are not accountable/ responsible for the patient's medical care. In contrast, physicians will remain responsible and accountable for the medical care that the patient receives.

11.5.4   INTELLECTUAL PROPERTY

A strong Intellectual Property System (IPS) is necessary to provide a platform for innovation. Government patent systems must offer adequate protection for innovations relating to personalized medicine, as well as conduction of patent examinations that will allow patents of appropriate scope and quality. The U.S. Patent and Trademark Office (PTO) have granted thousands of patents on human genes. A gene patent holder has the right to prevent anyone from studying, testing, or even looking at a gene's sequence. The AMA has well-established policy opposing the practice of PTO of issuing patents on human genes. A lawsuit was filed by Association for Molecular Pathology on behalf of researchers, genetic counselors, patients, cancer survivors, breast cancer and women's health groups, and scientific associations representing geneticists, physicians and laboratory professionals. The lawsuit was filed against the PTO, Myriad Genetics, and the University of Utah Research Foundation, which hold patents on the genes BRCA1 and BRCA2, mutations in which dramatically increase the risk for breast, ovarian and other cancers. The lawsuit charges that patents on human genes violate the first amendment and patent law because genes are "products of nature" and therefore cannot be patented.

In 2010, a federal court ruled that the patents on the BRCA1 and BRCA2 genes are invalid. The decision was reversed by a federal court of appeals. The Supreme Court will review the most recent decision by the court of appeals upholding gene patents. In a related case, Mayo v. Prometheus, the Supreme Court issued a unanimous decision invalidating patents on methods for evaluating a patient's response to a drug.

11.5.5 DATA MANAGEMENT AND ANALYSIS

In addition to many challenges facing physician practices in the process of adopting HIT and integrating it into their practice, personalized medicine is likely to present problem of data management and analysis challenges. Genetic testing, focusing on a few genes or a panel of genes, poses far smaller challenges than genomic testing which involves a large amount of data and a need for significant software support to conduct analysis and decision-support.

At present, there are no widely accepted genomic standards or quantitative performance metrics for confidence in variant calling when sequencing genomes. These are needed to achieve sound, reproducible research and regulated applications in physician clinical practice. A consortium has been launched by the National Institute of Standards and Technology to develop the reference materials, reference methods and reference data needed to assess confidence in human genome sequencing. The consortium's stated intention is to develop products that enable translation of whole genome sequencing to clinical applications by providing widely accepted materials, methods and data for performance assessment.

11.5.5.1   VARIANT DATABASES AND CLINICAL DECISION SUPPORT

Currently, early adopters of next-generation sequencing, including clinical whole genome analysis, have identified bioinformatics tools and clinical grade databases as a high-priority need. As with all data management and analysis, this raises questions about standards, interoperability, privacy and security. It is generally agreed that the clinical databases need to assure quality and facilitate data sharing. In addition to the clinical databases, there are many software tools under development that are resource intensive, but are not interoperable and do not follow any established standards. The FDA has begun to express an interest in regulation of such clinical decision-support software. This raises a host of questions for the role of professional medical specialty societies and the federal regulation of medicine. The National Institutes of Health (NIH) has established a genetic testing registry with the aim of providing to physicians and other interested stake holders information about every genetic test's purpose, methodology, validity, evidence of the test's usefulness, and laboratory contacts and credentials. The AMA has strongly supported this initiative in order to facilitate standardization and access by practicing clinicians to information on genetic testing.

11.5.5.2 COVERAGE AND PAYMENT

While a great deal of legislative and regulatory activity has focused on coding, coverage, and payment for genetic and genomic diagnostic tests and laboratory interpretive services, little attention has been directed towards assessing the impact on clinical practice once genetic and genomic information becomes integrated into clinical practice. While there could be significant savings associated with improved diagnosis, prognosis and targeted treatment, there is likely to be a marked increase in the time spent by treating physicians analyzing these additional test results, caring for patients' manifest conditions, and counseling patients. Diagnostic testing services, test kits and bioinformatics have enhanced the therapeutic output. The number of diagnostics in the personalized medicine arena has sharply risen within a very short period of time and the numbers are expected to grow. Presently more than 2500 genetic tests are available. Furthermore, tests based on next-generation sequencing technology resulting in a magnitude of data, must be processed for interpretation as computer software programs and complex multifactorial computations (bioinformatics and/or algorithms) that are likely to continuously change as we discover more information about the human genome. Many of these software programs and algorithms will have to be updated on a regular basis. However, it is difficult to forecast the extent to which there will be regular updates to diagnostics. Due to large volume of new tests, the complexity, and the lack of uniform standards governing the sequencing, bio informatics and algorithm portions of the tests, the regulatory approval and payment pathways are being taxed. The issue of value-based payments for molecular diagnostics and the genetic/genomic testing, lack of policies to cover proactive care are the big challenges for PM. Insufficient data on healthcare and economic outcomes for personalized medicine approaches often leads to failure of reimbursement decisions. Regionally fragmented Medicare reimbursement policy hinders adoption of genetic tests. Insurers generally do not recognize contribution of collateral services (laboratory services, genetic counseling, etc.) for equitable coverage of personalized medicine. The current reimbursement policy describes what is actually needed to ensure the physicians to apply personalized therapies. Personalized medicine challenges current assumptions made in coverage decisions. Compared to traditional diagnostics; certain imaging, molecular and genetic tests will provide a more precise information about present/future susceptibility to individual to any disease and response to treatment approach. Insurers typically do not provide coverage for tests of disease susceptibility.

11.5.5.3   REGULATION

A sound regulatory guideline is essential to nurture innovation in the field. All producers must have confidence in the process by which products are developed and regulated. Industry requires coordination and alignment among agencies for regulating diagnostic applications. They should be equipped with proper genetics specialty area. FDA requires more genomic data and input from industry for the development of new regulatory guidelines. The U.S. Food and Drug Administration (FDA) is playing a key role in identifying emerging trends, monitoring personalized medicine, taking decisions and providing guidelines for the implications of drug development and regulatory review.

11.5.5.4   LEGAL AND ETHICAL CONSIDERATIONS

The policy concerning the ethical considerations and professional obligations of physicians related to genetic testing is important. The fundamentals that run throughout existing opinions relating to personalized medicine include:
- well-informed, voluntary consent and appropriate counseling;
- protection of privacy/confidentiality to prevent unauthorized/inappropriate use of personal genetic information.

In short, these policies address general areas of concern, including informed consent, privacy and confidentiality associated with genetic information. To the extent that genetic or next-generation sequencing is undertaken in the clinical setting, privacy of results is covered by the Health Insurance Portability and Accountability Act. If the testing is strictly investigational, it is governed by the Common Rule (to the extent federal funds are involved in the research) that governs consent, privacy, confidentiality and disclosure to third parties. In contrast, there are no overarching federal or industry guidelines governing commercial genetic testing companies. They do not have to comply with any privacy controls, nor do they have limits on the use(s) of genetic data and information beyond those that are set by their own internal policies.

11.6   BARRIERS TO INTRODUCING PERSONALIZED MEDICINE INTO CLINICAL PRACTICE

Lack of training about genomic tests and interpretation of the results as an important barrier to use genomic testing in their practice. Lack of clinical guidelines and protocols on the use of genomic tests is a key issue. Physicians are interested to integrate genomics in their practice will need a lot more training and education to be able to know more about what is all about so how to counsel patients. Physicians were often concerned about complexities involved when interpreting and communicating genomic test results and the subsequent responsibilities that will add to their existing workloads.

They expressed concern about exploitation patient's genetic information by private companies (e.g., insurance companies, employers, etc.) in the absence of necessary regulations. Further, individuals might be discriminated against on the basis of this genetic information. Given increasing competition for health care resources they raised questions about the cost of genetic testing and whether they should be publicly funded to ensure equity in our health care system. These are all valid concerns that need to be addressed for successful integration of personalized medicine into clinical practice.

Establishment of setup for characterization of each and every individual's to start medication therapy is a tedious, costly and time taking work. Lack of funding sources, clear biomarkers or evidences to prove variability and disease, DNA sampling, characterization, storage and retrieval are costly processes and require well trained personnel and clinical study facilities, Regulatory guidelines that hamper smooth working, Due to diversity among individuals, different administrative bodies put forward different guidelines for dealing with data protection requirements.

11.7   FUTURE OF PERSONALIZED MEDICINE

Many physicians are agreeing that genetic developments will directly affect their medical practice in the future; however, they showed different viewpoints in terms of the timing of potential influences of genomics. Some believed that, considering the swift advances in genomics, personalized medicine would become an inevitable part of clinical practice. This argument was supported by the fact that some of the physicians had already received requests from their patients for interpretation of direct-to-consumer (DTC) genetic test results. These physicians suggested that technological advances in genomics have already started a gradual change in clinical practice. In contrast, some physicians believed that the effect of genomics on clinical practice may take more than 10 years to be realized.

11.8 CONCLUSIONS

Tailoring therapy based on pharmacogenomic testing results may save lives and improve patient care. Additionally, the decreasing price of new technologies that gather personal genomic information will facilitate their transition from basic research to bedside, thereby reshaping clinical diagnostic and therapeutic outcome. The challenges to healthcare professional include the enriching the knowledge of 'omics' like genomics, pharmacogenomic, and to consider how the new genomic information may be used to influence clinical decisions and to fulfill the promises of personalized healthcare.

KEYWORDS

- **Biomarkers**
- **Genetic testing**
- **Personalized medicine**
- **Pharmacogenetic**

REFERENCES

1. Carroll, K. M., & Nuro, K. F. (2002). One Size Cannot Fit All, A Stage Model for Psychotherapy Manual Development, Clinical Psychology, *Science and Practice, 9,* 396–406.
2. Amalia Issa M. (2007). "Personalized Medicine and the Practice of Medicine in the twenty-first century," *McGill Journal of Medicine, 10(1),* 53–57.
3. Brandi, M. L., Eric, A., Ann, D., Willard, D., Domnique, E., et al. (2012). Challenges Faced in the Integration of Pharmacogenetics/Genomics into Drug Development, *Journal of Pharmacogenomics & Pharmacoproteomics, 3,* 108.
4. Hunt, L. M., Truesdell, N. D., & Kreiner, M. J. (2013). Genes Ace, and Culture in Clinical Care, *Medical Anthropology Quarterly,* (DOI: 10.1111/maq.12026).
5. Mélanie, G., Gareau, Philip Sherman, M., & Allan Walker W. (2010). Probiotics and the Gut Microbiota in Intestinal Health and Disease, *Nature Reviews Gastroenterology and Hepatology, 7,* 503–514.

6. Johanne Tremblay, & Pavel Hamet (2013). Role of Genomics on the Path to Personalized Medicine, *62*, S2–S5

7. Sotiriou, C., & Piccart, M. J. (2007). Taking Gene-Expression Profiling to the Clinic, When Will Molecular Signatures Become Relevant to Patient Care? *Nat Rev Cancer, 7*, 545–553.

8. Vionnet, N., Stoffel, M., Takeda, J., Yasuda, K., Bell, G. I., Zouali, H. et al. (1992). Nonsense Mutation in the Glucokinase Gene Causes Early Onset Non-Insulin-Dependent Diabetes Mellitus, *Nature, 356*, 721, 722

9. Demenais, F., Kanninen, T., Lindgren, C. M., Wiltshire, S., Gaget, S., Dandrieux, C. et al. (2003). A Meta-Analysis of Four European Genome Screens (GIFT Consortium) Shows Evidence for a Novel Region on Chromosome 17p11.2–q22 Linked to Type 2 Diabetes, Hum Mol Genet, *12*, 1865–1873.

10. Limdi, N. A., & Veenstra, D. L. (2008). Warfarin Pharmaco Genetics, *Pharmacotherapy, 28*, 1084–1097.

11. Yuan, H. Y., Chen, J. J., Lee, M. T., Wung, J. C., Chen, Y. F., Charng, M. J. et al. (2005). A Novel Functional VKORC1 Promoter Polymorphism is Associated with Inter-Individual and Inter-Ethnic Differences in Warfarin Sensitivity, *Hum Mol Genet, 14*, 1745–1751.

12. Jiang, Y., & Wang, M. (2010). Personalized Medicine in Oncology Tailoring the Right Drug to the Right Patient, *Biomark Med, 4*, 523–533.

13. Link, E., Parish, S., Armitage, J., et al. (2008). SLCO1B1 Variants and Statininduced Myopathy—A Genomewide Study, *N Engl J Med, 359*, 789–799.

14. Ross, J. S., Slodkowska, E. A., Symmans, W. F., Pusztai, L., Ravdin, P. M., Hortobagyi, G. N. (2009). The HER-2 Receptor and Breast Cancer, Ten Years of Targeted Anti-HER-2 Therapy and Personalized Medicine, *Oncologist, 14*, 320–368.

15. Abramson, V., & Arteaga, C. L. (2011). New Strategies in HER2-Over Expressing Breast Cancer, Many Combinations of Targeted Drugs Available, *Clin Cancer Res, 17*, 952–958.

16. Jelovac, D., & Wolff, A. C. (2012). The Adjuvant Treatment of HER2-Positive Breast Cancer, *Curr Treat Options Oncol, 13*, 230–239.

17. Ross, J. S., Slodkowska, E. A., Symmans, W. F., Pusztai, L., Ravdin, P. M., & Hortobagyi, G. N. (2009). The HER-2 Receptor and Breast Cancer, Ten Years of Targeted Anti-HER-2 Therapy and Personalized Medicine, *Oncologist, 14*, 320–368.

18. Chung, W. H., Hung, S. I., Hong, H. S., Hsih, M. S., Yang, L. C., Ho, H. C., et al. (2004). Medical Genetics, a Marker for Stevense-Johnson Syndrome, *Nature, 428*, 486.

19. Chen, P., Lin, J. J., Lu, C. S., Ong, C. T., Hsieh, P. F., Yang, C. C. et al. (2011). Carbamazepine-Induced Toxic Effects and HLA-B*1502 Screening in Taiwan, *N Engl J Med, 364*, 1126–1133.

20. Locharernkul, C., Loplumlert, J., Limotai, C., Korkij, W., Desudchit, T., Tongkobpetch, S., et al. (2008). Carbamazepine and Phenytoin Induced Stevense-Johnson Syndrome is Associated with HLA-B*1502 Allele in Thai population, *Epilepsia, 49*, 2087–2091.

21. Kaniwa, N., Saito, Y., Aihara, M., Matsunaga, K., Tohkin, M., Kurose, K., et al. (2008). HLA-B Locus in Japanese Patients with Antiepileptics and Allopurinol-Related Stevense-Johnson Syndrome and Toxic Epidermal Necrolysis, *Pharmacogenomics, 9*, 1617–1622.

22. Chen, P., Lin, J. J., Lu, C. S., Ong, C. T., Hsieh, P. F, Yang, C. C. et al. (2011). Carbamaze pine-Induced Toxic Effects and HLA-B*1502 Screening in Taiwan N. Engl., *J Med, 364*, 1126–1133.

23. Shikata, E., Yamamoto, R., Takane, H., Shigemasa, C., Ikeda, T., Otsubo, K. et al. (2007). Human Organic Cation Transporter (OCT1 and OCT2) Gene Polymorphisms and Therapeutic Effects of Metformin, *J Hum Genet, 52*, 117–122.

24. Geng, X., Li, L., Bottino, R., Balamurugan, A. N., Bertera, S., Densmore, E. et al. (2007). Antidiabetic Sulfonylurea Stimulates Insulin Secretion Independently of Plasma Membrane KATP Channels, *Am J Physiol Endocrinol Metab, 293*, 293–301.
25. Aquilante, C. L. (2010). Sulfonylurea Pharmacogenomics in Type 2 Diabetes, the Influence of Drug Target and Diabetes Risk Polymorphisms, *Expert Rev Cardiovasc Ther, 8*, 359–372.
26. Pearson, E. R., Starkey, B. J., Powell, R. J., Gribble, F. M., Clark, P. M. & Hattersley, A. T. (2003). Genetic Cause of Hyper Glycaemia and Response to Treatment in Diabetes, *Lancet, 362*, 1275–1281.
27. Pearson, E. R., Donnelly, L. A., Kimber, C., Whitley, A., Doney, A. S., McCarthy, M. I., et al. (2007). Variation in TCF7L2 Influences Therapeutics Response to Sulfonylureas, a GoDARTs Study, *Diabetes, 56*, 2178–2182.
28. Pearson, E. R. (2009). Pharmacogenetics and Future Strategies in Treatings Hyperglycaemia in Diabetes, *Front Biosci, 14*, 4348–4362.

CHAPTER 12

SOME ASPECTS OF MOLECULAR MECHANICS AND DYNAMICS INTERACTIONS OF NANOSTRUCTURAL ELEMENTS

A. V. VAKHRUSHEV and A. M. LIPANOV

CONTENTS

ABSTRACT

In this chapter a systematic investigation on application of theoretical methods for calculation of internal structure and the equilibrium configuration (shape) of separate noninteracting nanoparticles by the molecular mechanics and dynamics interactions of nanostructural elements presented.

12.1   INTRODUCTION

The properties of a nanocomposite are determined by the structure and properties of the nanoelements, which form it. One of the main tasks in making nano composites is building the dependence of the structure and shape of the nanoelements forming the basis of the composite on their sizes. This is because with an increase or a decrease in the specific size of nanoelements (nano fibers, nanotubes, nanoparticles, etc.), their physical-mechanical properties such as coefficient of elasticity, strength, deformation parameter, etc. are varying over one order [1–5].

The calculations and experiments show that this is primarily due to a significant rearrangement (which is not necessarily monotonous) of the atomic structure and the shape of the nano element. The experimental investigation of the above parameters of the nanoelements is technically complicated and laborious because of their small sizes. In addition, the experimental results are often inconsistent. In particular, some authors have pointed to an increase in the distance between the atoms adjacent to the surface in contrast to the atoms inside the nano element, while others observe a decrease in the aforementioned distance [6].

Thus, further detailed systematic investigations of the problem with the use of theoretical methods, that is, mathematical modeling, are required.

The atomic structure and the shape of nanoelements depend both on their sizes and on the methods of obtaining which can be divided into two main groups:

1.   Obtaining nanoelements in the atomic coalescence process by "assembling" the atoms and by stopping the process when the nanoparticles grow to a desired size (the so-called "bottom-up" processes). The process of the particle growth is stopped by the change of physical or chemical conditions of the particle formation, by cutting off supplies of the substances that are necessary to form particles, or because of the limitations of the space where nanoelements form.

2.   Obtaining nanoelements by breaking or destruction of more massive (coarse) formations to the fragments of the desired size (the so-called "up down" processes).

In fact, there are many publications describing the modeling of the "bottom-up" processes [7, 8], while the "up down" processes have been studied very little. Therefore, the objective of this work is the investigation of the regularities of the changes in the structure and shape of nanoparticles formed in the destruction ("up down") processes depending on the nanoparticle sizes, and building up theoretical dependences describing the above parameters of nanoparticles.

When the characteristics of powder nano composites are calculated it is also very important to take into account the interaction of the nanoelements since the changes in their original shapes and sizes in the interaction process and during the formation (or usage) of the nanocomposite can lead to a significant change in its properties and a cardinal structural rearrangement. In addition, the experimental investigations show the appearance of the processes of ordering and self-assembling leading to a more organized form of a nano system [9–15]. In general, three main processes can be distinguished: the first process is due to the regular structure formation at the interaction of the nano structural elements with the surface where they are situated; the second one arises from the interaction of the nano structural elements with one another; the third process takes place because of the influence of the ambient medium surrounding the nano structural elements. The ambient medium influence can have "isotropic distribution" in the space or it can be presented by the action of separate active molecules connecting nanoelements to one another in a certain order. The external action significantly changes the original shape of the structures formed by the nanoelements. For example, the application of the external tensile stress leads to the "stretch" of the nano element system in the direction of the maximal tensile stress action; the rise in temperature, vice versa, promotes a decrease in the spatial anisotropy of the nanostructures [10]. Note that in the self-organizing process, parallel with the linear moving, the nanoelements are in rotary movement. The latter can be explained by the action of moment of forces caused by the asymmetry of the interaction force fields of the nanoelements, by the presence of the "attraction" and "repulsion" local regions on the nano element surface, and by the "nonisotropic" action of the ambient as well.

The above phenomena play an important role in nano technological processes. They allow developing nanotechnologies for the formation of nanostructures by the self-assembling method (which is based on self-organizing processes) and building up complex spatial nanostructures consisting of different nanoelements (nanoparticles, nanotubes, fullerenes, super-molecules, etc.) [15]. However, in a number of cases, the tendency towards self-organization interferes with the formation of a desired nanostructure. Thus, the nanostructure arising from the self-organizing process is, as a rule, "rigid" and stable against external actions. For example, the "adhesion" of nanoparticles interferes with the use of separate nanoparticles in various nano technological processes, the uniform mixing of the nanoparticles from different materials and the formation of nanocomposite with desired properties. In connection with this, it is important to model the processes of static and dynamic interaction of the nanostructure elements. In this case, it is essential to take into consideration the interaction force moments of the nanostructure elements, which causes the mutual rotation of the nanoelements.

The investigation of the above dependences based on the mathematical modeling methods requires the solution of the aforementioned problem on the atomic level. This requires large computational aids and computational time, which makes the development of economical calculation methods urgent. The objective of this work was the development of such a technique.

This chapter gives results of the studies of problems of numeric modeling within the framework of molecular mechanics and dynamics for investigating the regularities of the amorphous phase formation and the nucleation and spread of the crystalline or hypo crystalline phases over the entire nanoparticle volume depending on the process parameters, nanoparticles sizes and thermodynamic conditions of the ambient. Also the method for calculating the interactions of nano structural elements is offered, which is based on the potential built up with the help of the approximation of the numerical calculation results using the method of molecular dynamics of the pair wise static interaction of nanoparticles. Based on the potential of the pair wise interaction of the nanostructure elements, which takes into account forces and moments of forces, the method for calculating the ordering and self-organizing processes has been developed. The investigation results on the self-organization of the system consisting of two or more particles are presented and the analysis of the equilibrium stability of various types of nanostructures has been carried out. These results are a generalization of the authors' research in Refs. [16–24]. A more detailed description of the problem you can obtain in these works.

12.2   PROBLEM STATEMENT

The problem on calculating the internal structure and the equilibrium configuration (shape) of separate noninteracting nanoparticles by the molecular mechanics and dynamics methods has two main stages:

1. The "initiation" of the task, that is, the determination of the conditions under which the process of the nanoparticle shape and structure formation begins.

2. The process of the nanoparticle formation.

Note that the original coordinates and initial velocities of the nanoparticle atoms should be determined from the calculation of the macroscopic parameters of the destructive processes at static and dynamic loadings taking place both on the nano-scale and on the macroscale. Therefore, in the general case, the coordinates and velocities are the result of solving the problem of modeling physical-mechanical destruction processes at different structural levels. This problem due to its enormity and complexity is not considered in this paper. The detailed description of its statement and the numerical results of its solution are given in the works of the authors [16–19].

The problem of calculating the interaction of ordering and self-organization of the nanostructure elements includes three main stages: the first stage is building the internal structure and the equilibrium configuration (shape) of each separate noninteracting nanostructure element; the second stage is calculating the pair wise interaction of two nanostructure elements; and the third stage is establishing the regularities of the spatial structure and evolution with time of the nanostructure as a whole.

Let us consider the above problems in sequence.

12.2.1 THE CALCULATION OF THE INTERNAL STRUCTURE AND THE SHAPE OF THE NON-INTERACTING NANOELEMENTS

The initialization of the problem is in giving the initial coordinates and velocities of the nanoparticle atoms

$$\vec{\mathbf{x}}_i = \vec{\mathbf{x}}_{i0}, \vec{\mathbf{V}}_i = \vec{\mathbf{V}}_{i0}, \ t = 0, \ \vec{\mathbf{x}}_i \subset \Omega_k \,, \tag{1}$$

Where $\vec{\mathbf{x}}_{i0}, \vec{\mathbf{x}}_i$ are original and current coordinates of the i-th atom; $\vec{\mathbf{v}}_{i0}, \vec{\mathbf{v}}_i$ are initial and current velocities of the i-th atom, respectively; Ω_k is an area occupied by the nano element.

The problem of calculating the structure and the equilibrium configuration of the nano element will be carried out with the use of the molecular dynamics method taking into consideration the interaction of all the atoms forming the nano element. Since, at the first stage of the solution, the nano element is not exposed to the action of external forces, it is taking the equilibrium configuration with time, which is further used for the next stage of calculations.

At the first stage, the movement of the atoms forming the nanoparticle is determined by the set of Langev in differential equations at the boundary conditions (1) [25]

$$m_i \cdot \frac{d\vec{V}_i}{dt} = \sum_{j=1}^{N_k} \vec{F}_{ij} + \vec{F}_i(t) - \alpha_i m_i \vec{V}_i, \qquad i = 1, 2, .., N_k,$$

$$\frac{d\vec{x}_i}{dt} = \vec{V}_i, \tag{2}$$

where N_k is the number of atoms forming each nanoparticle; m_i is the mass of the i-th atom; α_i is the "friction" coefficient in the atomic structure; $\vec{F}_i(t)$ is a random set of forces at a given temperature which is given by Gaussian distribution.

The inter atomic interaction forces usually are potential and determined by the relation

$$\vec{F}_{ij} = -\sum_{1}^{n} \frac{\partial \Phi(\vec{\rho}_{ij})}{\partial \vec{\rho}_{ij}} , \quad i = 1, 2, ..., N_k, \quad j = 1, 2, ..., N_k, \tag{3}$$

where $\vec{\rho}_{ij}$ is a radius-vector determining the position of the i-th atom relative to the j-th atom; $\Phi(\vec{\rho}_{ij})$ is a potential depending on the mutual positions of all the atoms; n is the number of inter atomic interaction types.

In the general case, the potential $\Phi(\vec{\rho}_{ij})$ is given in the form of the sum of several components corresponding to different interaction types:

$$\Phi(\vec{\rho}_{ij}) = \Phi_{cb} + \Phi_{va} + \Phi_{ta} + \Phi_{pg} + \Phi_{vv} + \Phi_{es} + \Phi_{hb}. \tag{4}$$

Here the following potentials are implied: Φ_{cb}—of chemical bonds; Φ_{va}—of valence angles; Φ_{ta}—of torsion angles; Φ_{pg}—of flat groups; Φ_{vv}—of Vander Waals contacts; Φ_{es}—of electrostatics; Φ_{hb}—of hydrogen bonds.

The above addends have different functional forms. The parameter values for the interaction potentials are determined based on the experiments (crystallography, spectral, calorimetric, etc.) and quantum calculations [25].

Giving original coordinates (and forces of atomic interactions) and velocities of all the atoms of each nanoparticle in accordance with Eq. (2), at the start time, we find the change of the coordinates and the velocities of each nanoparticle atoms with time from the equation of motion (1). Since the nanoparticles are not exposed to the action of external forces, they take some atomic equilibrium configuration with time that we will use for the next calculation stage.

12.2.2 THE CALCULATION OF THE PAIR WISE INTERACTION OF THE TWO NANOSTRUCTURE ELEMENTS

At this stage of solving the problem, we consider two interacting nanoelements. First, let us consider the problem statement for symmetric nanoelements, and then for arbitrary shaped nanoelements.

First of all, let us consider two symmetric nanoelements situated at the distance S from one another (Fig. 12.1) at the initial conditions

$$\vec{x}_i = \vec{x}_{i0}, \vec{V}_i = 0. \ t = 0, \vec{x}_i \subset \Omega_1 \bigcup \Omega_2 \ , \tag{5}$$

where Ω_1, Ω_2 are the areas occupied by the first and the second nanoparticle, respectively.

We obtain the coordinates $\vec{x}_{i0}$ from Eq. (2) solution at initial conditions (1). It allows calculating the combined interaction forces of the nanoelements

$$\vec{F}_{b1} = -\vec{F}_{b2} = \sum_{i-1}^{N_1} \sum_{j=1}^{N_2} \vec{F}_{ij} \ , \tag{6}$$

where $i, j,$ are the atoms and N_1, N_2 are the numbers of atoms in the first and in the second nanoparticle, respectively.

Forces $\vec{F}_{ij}$ are defined from Eq. (3).

In the general case, the force magnitude of the nanoparticle interaction $\left| \vec{F}_{bi} \right|$ can be written as product of functions depending on the sizes of the nanoelements and the distance between them:

$$\left| \vec{F}_{bi} \right| = \Phi_{11}(S_c) \times \Phi_{12}(D) \tag{7}$$

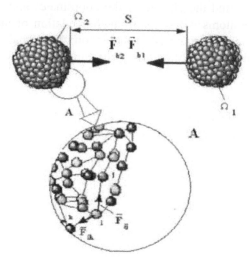

FIGURE 12.1   The schemes of the nanoparticle interaction; A–an enlarged view of the nanoparticle fragment.

The $\vec{F}_{bi}$ vector direction is determined by the direction cosines of a vector connecting the centers of the nanoelements.

Now, let us consider two interacting asymmetric nanoelements situated at the distance S_c between their centers of mass (Fig. 12.2) and oriented at certain specified angles relative to each other.

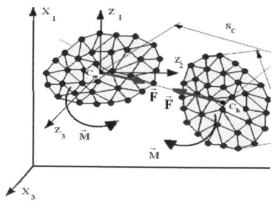

FIGURE 12.2   Two interacting nanoelements; $\vec{M}$, $\vec{F}$ are the principal moment and the principal vector of the forces, respectively.

In contrast to the previous problem, inter atomic interaction of the nano-elements leads not only to the relative displacement of the nanoelements but to their rotation as well. Consequently, in the general case, the sum of all the forces of the inter atomic interactions of the nanoelements is brought to the principal vector of forces $\vec{F}_c$ and the principal moment $\vec{M}_c$

$$\vec{F}_c = \vec{F}_{b1} = -\vec{F}_{b2} = \sum_{i=1}^{N_1} \sum_{j=1}^{N_2} \vec{F}_{ij} \, , \tag{8}$$

$$\vec{M}_c = \vec{M}_{c1} = -\vec{M}_{c2} = \sum_{i=1}^{N_1} \sum_{j=1}^{N_2} \vec{\rho}_{cj} \times \vec{F}_{ij} \, , \tag{9}$$

where $\vec{\rho}_{cj}$ is a vector connecting points c and j.

The main objective of this calculation stage is building the dependences of the forces and moments of the nanostructure nano element interactions on the distance S_c between the centers of mass of the nanostructure nanoelements, on the angles of mutual orientation of the nanoelements $\Theta_1, \Theta_2, \Theta_3$ (shapes of the nanoelements) and on the characteristic size D of the nano element. In the general case, these dependences can be given in the form.

$$\vec{F}_{bi} = \vec{\Phi}_F(S_c, \Theta_1, \Theta_2, \Theta_3, D) \, , \tag{10}$$

$$\vec{M}_{bi} = \vec{\Phi}_M(S_c, \Theta_1, \Theta_2, \Theta_3, D) \, , \tag{11}$$

For spherical nanoelements, the angles of the mutual orientation do not influence the force of their interaction; therefore, in Eq. (12), the moment is zero.

In the general case, Eqs. (11) and (12) can be approximated by analogy with Eq. (8), as the product of functions $S_0, \Theta_1, \Theta_2, \Theta_3, D$, respectively. For the further numerical solution of the problem of the self-organization of nanoele-ments, it is sufficient to give the above functions in their tabular form and to use the linear (or nonlinear) interpolation of them in space.

12.3   PROBLEM FORMULATION FOR INTERACTION OF SEVERAL NANOELEMENTS

When the evolution of the nano system as whole (including the processes of ordering and self-organization of the nanostructure nanoelements) is investi-gated, the movement of each system nano element is considered as the movement of a single whole. In this case, the translational motion of the center of mass of each nano element is given in the coordinate system X_1, X_2, X_3, and the nano element rotation is described in the coordinate system Z_1, Z_2, Z_3, which is related to the center of mass of the nano element (Fig. 12.2). The system of equations describing the above processes has the form

$$
\left[
\begin{aligned}
M_k \frac{d^2 X_1^k}{dt^2} &= \sum_{j=1}^{N_e} F_{X_1}^{kj} + F_{X_1}^{ke}, \\[2mm]
M_k \frac{d^2 X_2^k}{dt^2} &= \sum_{j=1}^{N_e} F_{X_2}^{kj} + F_{X_2}^{ke}, \\[2mm]
M_k \frac{d^2 X_3^k}{dt^2} &= \sum_{j=1}^{N_e} F_{X_3}^{kj} + F_{X_3}^{ke}, \\[2mm]
J_{Z_1}^k \frac{d^2 \Theta_1^k}{dt^2} + \frac{d\Theta_2^k}{dt} \cdot \frac{d\Theta_3^k}{dt}(J_{Z_3}^k - J_{Z_2}^k) &= \sum_{j=1}^{N_e} M_{Z_1}^{kj} + M_{Z_1}^{ke}, \\[2mm]
J_{Z2}^k \frac{d^2 \Theta_2^k}{dt^2} + \frac{d\Theta_1^k}{dt} \cdot \frac{d\Theta_3^k}{dt}(J_{Z_1}^k - J_{Z_3}^k) &= \sum_{j=1}^{N_e} M_{Z_2}^{kj} + M_{Z_2}^{ke}, \\[2mm]
J_{Z_3}^k \frac{d^2 \Theta_3^k}{dt^2} + \frac{d\Theta_2^k}{dt} \cdot \frac{d\Theta_1^k}{dt}(J_{Z_2}^k - J_{Z_1}^k) &= \sum_{j=1}^{N_e} M_{Z_3}^{kj} + M_{Z_3}^{ke},
\end{aligned}
\right.
\tag{12}
$$

where x_i^k, Θ_i^k are coordinates of the centers of mass and angles of the spatial orientation of the principal axes Z_1, Z_2, Z_3 of nanoelements; $F_{X_1}^{kj}, F_{X_2}^{kj}, F_{X_3}^{kj}$ are the interaction forces of nanoelements; $F_{X_1}^{ke}, F_{X_2}^{ke}, F_{X_3}^{ke}$ are external forces acting on nanoelements; N_e is the number of nanoelements; M_k is a mass of a nano element; $M_{X_1}^{kj}, M_{X_2}^{kj}, M_{X_3}^{kj}$ is the moment of forces of the nano element interaction; $M_{X_1}^{ke}, M_{X_2}^{ke}, M_{X_3}^{ke}$ are external moments acting on nanoelements; $J_{Z_1}^k, J_{Z_2}^k, J_{Z_3}^k$ are moments of inertia of a nano element.

The initial conditions for the system of Eqs. (13) and (14) have the form

$$
\vec{X}^k = \vec{X}_0^k; \quad \Theta^k = \Theta_0^k; \quad \vec{V}^k = \vec{V}_0^k; \quad \frac{d\Theta^k}{dt} = \frac{d\Theta_0^k}{dt}; \quad t = 0,
\tag{13}
$$

12.4 NUMERICAL PROCEDURES AND SIMULATION TECHNIQUES

In the general case, the problem formulated in the previous sections has no analytical solution at each stage; therefore, numerical methods for solving are used, as a rule. In this work, for the first stages, the numerical integration of the equation of motion of the nanoparticle atoms in the relaxation process are used in accordance with Verlet scheme [26]:

$$\vec{x}_i^{n+1} = \vec{x}_i^n + \Delta t\, \vec{V}_i^n + \left((\Delta t)^2 / 2m_i\right)\left(\sum_{j=1}^{N_k} \vec{F}_{ij} + \vec{F}_i - \alpha_i m_i \vec{V}_i^n\right)^n, \tag{14}$$

$$\vec{V}_i^{n+1} = (1 - \Delta t \alpha_i)\vec{V}_i^n + (\Delta t / 2m_i)((\sum_{j=1}^{N_k} \vec{F}_{ij} + \vec{F}_i)^n + (\sum_{j=1}^{N_k} \vec{F}_{ij} + \vec{F}_i)^{n+1}), \tag{15}$$

where $\vec{x}_i^n$, $\vec{V}_i^n$ are a coordinate and a velocity of the i-th atom at the n-th step with respect to the time; Δt is a step with respect to the time.

The solution of the Eq. (13) also requires the application of numerical methods of integration. In the present work, Runge-Kutta method [27] is used for solving Eq. (13).

$$(X_i^k)_{n+1} = (X_i^k)_n + (V_i^k)_n \Delta t + \frac{1}{6}(\mu_{1i}^k + \mu_{2i}^k + \mu_{3i}^k)\Delta t, \tag{16}$$

$$(V_i^k)_{n+1} = (V_i^k)_n + \frac{1}{6}(\mu_{1i}^k + 2\mu_{2i}^k + 2\mu_{3i}^k + \mu_{4i}^k). \tag{17}$$

$$\mu_{1i}^k = \Phi_i^k(t_n; (X_i^k)_n, \dots; (V_i^k)_n \dots)\Delta t,$$

$$\mu_{2i}^k = \Phi_i^k(t_n + \frac{\Delta t}{2}; (X_i^k + V_i^k \frac{\Delta t}{2})_n, \dots; (V_i^k)_n + \frac{\mu_{1i}^k}{2}, \dots)\Delta t,$$

$$\mu_{3i}^k = \Phi_i^k(t_n + \frac{\Delta t}{2}; (X_i^k + V_i^k \frac{\Delta t}{2} + \mu_{1i}^k \frac{\Delta t}{4})_n, \dots; (V_i^k)_n + \frac{\mu_{2i}^k}{2}, \dots)\Delta t, \tag{18}$$

$$\mu_{4i}^k = \Phi_i^k(t_n + \Delta t; (X_i^k + V_i^k \Delta t + \mu_{2i}^k \frac{\Delta t}{2})_n, \dots; (V_i^k)_n + \mu_{2i}^k, \dots)\Delta t.$$

$$\Phi_i^k = \frac{1}{M_k}(\sum_{j=1}^{N_e} F_{X_3}^{kj} + F_{X_3}^{ke}) \tag{19}$$

$$(\Theta_i^k)_{n+1} = (\Theta_i^k)_n + (\frac{d\Theta_i^k}{dt})_n \Delta t + \frac{1}{6}(\lambda_{1i}^k + \lambda_{2i}^k + \lambda_{3i}^k)\Delta t \tag{20}$$

$$(\frac{d\Theta_i^k}{dt})_{n+1} = (\frac{d\Theta_i^k}{dt})_n + \frac{1}{6}(\lambda_{1i}^k + 2\lambda_{2i}^k + 2\lambda_{3i}^k + \lambda_{4i}^k) \tag{21}$$

$$\lambda_{1i}^k = \Psi_i^k(t_n; (\Theta_i^k)_n, \ldots; (\frac{d\Theta_i^k}{dt})_n, \ldots)\Delta t \ ,$$

$$\lambda_{2i}^k = \Psi_i^k(t_n + \frac{\Delta t}{2}; (\Theta_i^k + \frac{d\Theta_i^k}{dt}\frac{\Delta t}{2})_n, \ldots; (\frac{d\Theta_i^k}{dt})_n + \frac{\lambda_{1i}^k}{2}, \ldots)\Delta t \ ,$$

$$\lambda_{3i}^k = \Psi_i^k(t_n + \frac{\Delta t}{2}; (\Theta_i^k + \frac{d\Theta_i^k}{dt}\frac{\Delta t}{2} + \lambda_{1i}^k\frac{\Delta t}{4})_n, \ldots; (\frac{d\Theta_i^k}{dt})_n + \frac{\lambda_{2i}^k}{2}, \ldots)\Delta t \ , \tag{22}$$

$$\lambda_{4i}^k = \Psi_i^k(t_n + \Delta t; (\Theta_i^k + \frac{d\Theta_i^k}{dt}\Delta t + \lambda_{2i}^k\frac{\Delta t}{2})_n, \ldots; (\frac{d\Theta_i^k}{dt})_n + \lambda_{2i}^k, \ldots)\Delta t \ .$$

$$\Psi_1^k = \frac{1}{J_{Z_1}^k}(-\frac{d\Theta_2^k}{dt} \cdot \frac{d\Theta_3^k}{dt}(J_{Z_3}^k - J_{Z_2}^k) + \sum_{j=1}^{N_e} M_{Z_1}^{kj} + M_{Z_1}^{ke}) \ ,$$

$$\Psi_2^k = \frac{1}{J_{Z_2}^k}(-\frac{d\Theta_1^k}{dt} \cdot \frac{d\Theta_3^k}{dt}(J_{Z_1}^k - J_{Z_3}^k) + \sum_{j=1}^{N_e} M_{Z_2}^{kj} + M_{Z_2}^{ke}) \ , \tag{23a}$$

$$\Psi_3^k = \frac{1}{J_{Z_3}^k}(-\frac{d\Theta_1^k}{dt} \cdot \frac{d\Theta_2^k}{dt}(J_{Z_1}^k - J_{Z_2}^k) + \sum_{j=1}^{N_e} M_{Z_3}^{kj} + M_{Z_3}^{ke})$$

were $i = 1,2,3$; $k = 1,2,\ldots N_e$

12.5   RESULTS AND DISCUSSIONS

Let us consider the realization of the above procedure taking as an example the calculation of the metal nanoparticle.

The potentials of the atomic interaction of Morse (23b) and Lennard-Johns (24) were used in the following calculations

$$\Phi(\bar{\rho}_{ij})_m = D_m(\exp(-2\lambda_m(|\bar{\rho}_{ij}| - -\rho_0)) - 2\exp(-\lambda_m(|\bar{\rho}_{ij}| - \rho_0))) \ , \tag{23b}$$

$$\Phi(\bar{\rho}_{ij})_{LD} = 4\varepsilon\left[\left(\frac{\sigma}{|\bar{\rho}_{ij}|}\right)^{12} - \left(\frac{\sigma}{|\bar{\rho}_{ij}|}\right)^{6}\right] \ , \tag{24}$$

where $D_m, \lambda_m, \rho_0, \varepsilon, \sigma$ are the constants of the materials studied.

For sequential and parallel solving the molecular dynamics equations, the program package developed at Applied Mechanics Institute, the Ural Branch of the Russian Academy of Sciences, and the advanced program package NAMD developed at the University of Illinois and Beckman Institute (USA) by the Theoretical Biophysics Group were used. The graphic imaging of the nanoparticle calculation results was carried out with the use of the program package VMD.

12.5.1 STRUCTURE AND FORMS OF NANOPARTICLES

At the first stage of the problem, the coordinates of the atoms positioned at the ordinary material lattice points (Fig. 12.3a) were taken as the original coordinates. During the relaxation process, the initial atomic system is rearranged into a new "equilibrium" configuration (Fig. 12.3b) in accordance with the calculations based on Eqs. (6)–(9), which satisfies the condition when the system potential energy is approaching the minimum (Fig. 12.3, the plot).

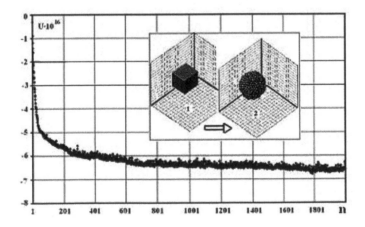

FIGURE 12.3 The initial crystalline (a) and cluster (b) structures of the nanoparticle consisting of 1331 atoms after relaxation; the plot of the potential energy U [J] variations for this atomic system in the relaxation process (*n*-number of iterations with respect to the time)

After the relaxation, the nanoparticles can have quite diverse shapes: globe-like, spherical centered, spherical eccentric, spherical icosahedral nanoparticles and asymmetric nanoparticles (Fig. 12.4).

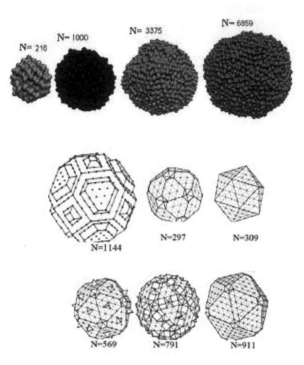

FIGURE 12.4 Nanoparticles of diverse shapes, depending on the number of atoms they consist of.

In this case, the number of atoms N significantly determines the shape of a nanoparticle. Note, that symmetric nanoparticles are formed only at a certain number of atoms. As a rule, in the general case, the nanoparticle deviates from the symmetric shape in the form of irregular raised portions on the surface. Besides, there are several different equilibrium shapes for the same number of atoms. The plot of the nanoparticle potential energy change in the relaxation process (Fig. 12.5) illustrates it.

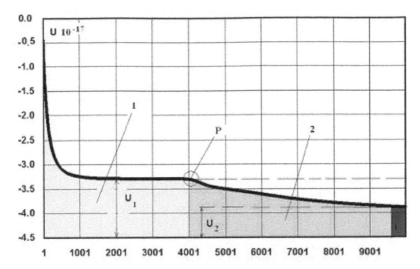

FIGURE 12.5 The plot of the potential energy changes of the nanoparticle in the relaxation process. 1 – a region of the stabilization of the first nanoparticle equilibrium shape; 2 – a region of the stabilization of the second nanoparticle equilibrium shape; P – a region of the transition of the first nanoparticle equilibrium shape into the second one.

As it follows from Fig. 12.5, the curve has two areas: the area of the decrease of the potential energy and the area of its stabilization promoting the formation of the first nanoparticle equilibrium shape (1). Then, a repeated decrease in the nanoparticle potential energy and the stabilization area corresponding to the formation of the second nanoparticle equilibrium shape are observed (2). Between them, there is a region of the transition from the first shape to the second one (P). The second equilibrium shape is more stable due to the lesser nanoparticle potential energy. However, the first equilibrium shape also "exists" rather long in the calculation process. The change of the equilibrium shapes is especially characteristic of the nanoparticles with an "irregular" shape. The internal structure of the nanoparticles is of importance since their atomic structure significantly differs from the crystalline structure of the bulk materials: the distance between the atoms and the angles change, and the surface formations of different types appear. In Fig. 12.6, the change of the structure of a two-dimensional nanoparticle in the relaxation process is shown.

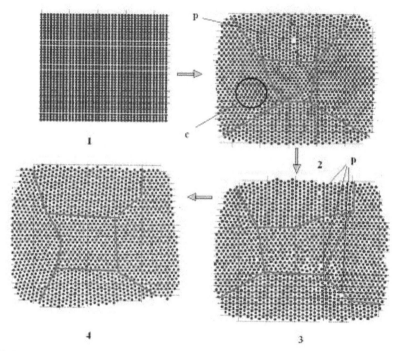

FIGURE 12.6 The change of the structure of a two-dimensional nanoparticle in the relaxation process: 1 – the initial crystalline structure; 2, 3, 4 – the nanoparticles structures which change in the relaxation process; p – pores; c – the region of compression.

The figure shows how the initial nanoparticle crystalline structure (1) is successively rearranging with time in the relaxation process (positions 2, 3, 4). Note that the resultant shape of the nanoparticle is not round, i.e., it has "remembered" the initial atomic structure. It is also of interest that in the relaxation process, in the nanoparticle, the defects in the form of pores (designation "p" in the figure) and the density fluctuation regions (designation "c" in the figure) have been formed, which are absent in the final structure.

12.5.2 NANOPARTICLES INTERACTION

Let us consider some examples of nanoparticles interaction. Figure 12.7 shows the calculation results demonstrating the influence of the sizes of the nanoparticles on their interaction force. One can see from the plot that the larger nanoparticles are attracted stronger, that is, the maximal interaction force increases with the size growth of the particle. Let us divide the interaction force of the nanoparticles by its maximal value for each nanoparticle size, respectively. The obtained plot of the "relative" (dimensionless) force (Fig. 12.8) shows that the value does not practically depend on the nanoparticle size since all the curves come close and can be approximated to one line.

Figure 12.9 displays the dependence of the maximal attraction force between the nanoparticles on their diameter that is characterized by nonlinearity and a general tendency towards the growth of the maximal force with the nanoparticle size growth.

The total force of the interaction between the nanoparticles is determined by multiplying of the two plots (Figs. 12.8 and 12.9).

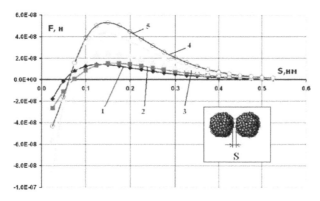

FIGURE 12.7   The dependence of the interaction force F [N] of the nanoparticles on the distance S [nm] between them and on the nanoparticle size: $1 - d = 2.04$; $2 - d = 2.40$; $3 - d = 3.05$; $4 - d = 3.69$; $5 - d = 4.09$ [nm].

Using the polynomial approximation of the curve in Fig. 12.5 and the power mode approximation of the curve in Fig. 12.6, we obtain

$$\bar{F} = (-1.13S^6 + 3.08S^5 - 3.41S^4 - 0.58S^3 + 0.82S - 0.00335)10^3 , \tag{25}$$

$$F_{max}{}^{\cdot} = 0.5 \cdot 10^{-9} \cdot d^{1.499} , \tag{26}$$

$$F = F_{max} \cdot \bar{F} , \tag{27}$$

where d and S are the diameter of the nanoparticles and the distance between them [nm], respectively; F_{max} is the maximal force of the interaction of the nanoparticles [N].

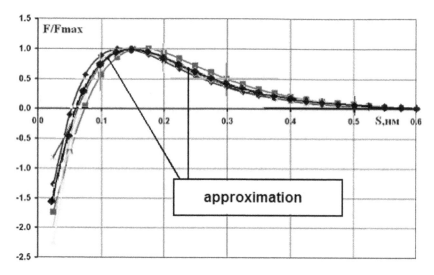

FIGURE 12.8   The dependence of the "relative" force $\overline{F}$ of the interaction of the nanoparticles on the distances S[nm] between them.

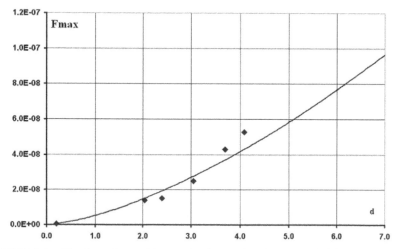

FIGURE 12.9   The dependence of the maximal attraction force F_{max} [N] the nanoparticles on the nanoparticle diameter d [nm].

Dependences (25)–(27) were used for the calculation of the nanocomposite ultimate strength for different patterns of nanoparticles' "packing" in the composite (Fig. 12.10).

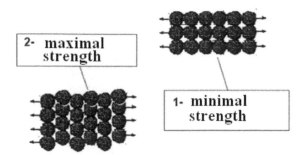

FIGURE 12.10   Different types of the nanoparticles' "packing" in the composite.

Figure 12.11 shows the dependence of the ultimate strength of the nanocomposite formed by mono disperse nanoparticles on the nanoparticle sizes. One can see that with the decrease of the nanoparticle sizes, the ultimate strength of the nano material increases, and vice versa. The calculations have shown that the nanocomposite strength properties are significantly influenced by the nanoparticles' "packing" type in the material. The material strength grows when the packing density of nanoparticles increases. It should be specially noted that the material strength changes in inverse proportion to the nanoparticle diameter in the degree of 0.5, which agrees with the experimentally established law of strength change of nanomaterials (the law by Hall-Petch) [18]:

$$\sigma = C \cdot d^{-0.5} , \tag{28}$$

where $C = C_{max} = 2.17 \times 10^4$ is for the maximal packing density; $C = C_{min} = 6.4 \times 10^3$ is for the minimal packing density.

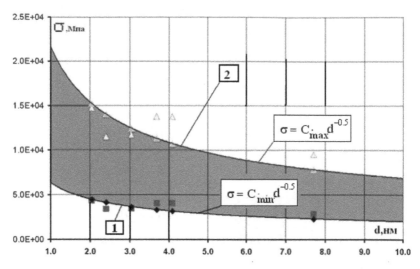

FIGURE 12.11   The dependence of the ultimate strength σ [MPa] of the nanocomposite formed by monodisperse nanoparticles on the nanoparticle sizes d [nm].

The electrostatic forces can strongly change force of interaction of nanoparticles. For example, numerical simulation of charged sodium (NaCl) nanoparticles system (Fig. 12.12) has been carried out. Considered ensemble consists of eight separate nanoparticles. The nanoparticles interact due to Vander-Waals and electrostatic forces.

Results of particles center of masses motion are introduced at (Fig.12.13) representing trajectories of all nanoparticles included into system. It shows the dependence of the modulus of displacement vector $|R|$ on time. One can see that nanoparticle moves intensively at first stage of calculation process. At the end of numerical calculation, all particles have got new stable locations, and the graphs of the radius vector $|R|$ become stationary. However, the nanoparticles continue to "vibrate" even at the final stage of numerical calculations. Nevertheless, despite of "vibration," the system of nanoparticles occupies steady position.

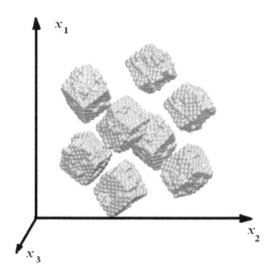

FIGURE 12.12 Nanoparticles system consists of eight nanoparticles NaCl.

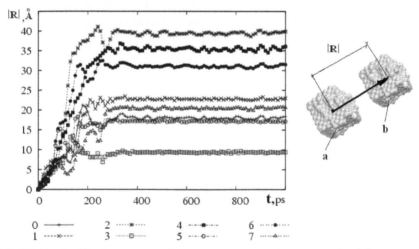

FIGURE 12.13 The dependence of nanoparticle centers of masses motion $|R|$ on the time t; a, b-the nanoparticle positions at time 0 and t, accordingly; 1–8 are the numbers of the nanoparticles.

However, one can observe a number of other situations. Let us consider, for example, the self-organization calculation for the system consisting of 125cubic nanoparticles, the atomic interaction of which is determined by Morse potential (Fig. 12.14).

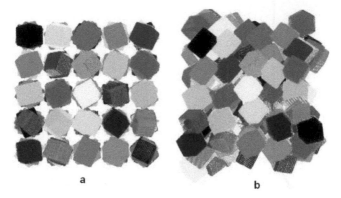

FIGURE 12.14   The positions of the 125cubic nanoparticles: a–initial configuration; b–final configuration of nanoparticles.

As you see, the nanoparticles are moving and rotating in the self-organization process forming the structure with minimal potential energy. Let us consider, for example, the calculation of the self-organization of the system consisting of two cubic nanoparticles, the atomic interaction of which is determined by Morse potential [12]. Figure 12.15 displays possible mutual positions of these nanoparticles. The positions, where the principal moment of forces is zero, corresponds to pairs of the nanoparticles 2, 3; 3, 4; 2–5 (Fig. 12.15) and defines the possible positions of their equilibrium.

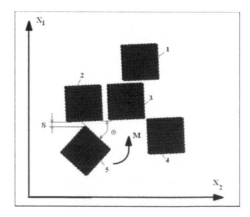

FIGURE 12.15   Characteristic positions of the cubic nanoparticles.

Figure 12.16 presents the dependence of the moment of the interaction force between the cubic nanoparticles 1–3 (Fig. 12.15) on the angle of their relative rotation. From the plot follows that when the rotation angle of particle 1 relative to particle 3 is $\pi/4$, the force moment of their interaction is zero. At an increase or a decrease in the angle the force moment appears. In the range of $\pi/8 < \theta < 3\pi/4$ the moment is small. The force moment rapidly grows outside of this range. The distance S between the nanoparticles plays a significant role in establishing their equilibrium. If $S > S_0$ (where S_0 is the distance, where the interaction forces of the nanoparticles are zero), then the particles are attracted to one another. In this case, the sign of the moment corresponds to the sign of the angle θ deviation from $\pi/4$. At $S < S_0$ (the repulsion of the nanoparticles), the sign of the moment is opposite to the sign of the angle deviation. In other words, in the first case, the increase of the angle deviation causes the increase of the moment promoting the movement of the nano element in the given direction, and in the second case, the angle deviation causes the increase of the moment hindering the movement of the nano element in the given direction. Thus, the first case corresponds to the unstable equilibrium of nanoparticles, and the second case to their stable equilibrium. The potential energy change plots for the system of the interaction of two cubic nanoparticles (Fig. 12.17) illustrate the influence of the parameter S. Here, curve 1 corresponds to the condition $S < S_0$ and it has a well-expressed minimum in the $0.3 < \theta < 1.3$ region. At $\theta < 0.3$ and $\theta > 1.3$, the interaction potential energy sharply increases, which leads to the return of the system into the initial equilibrium position. At $S > S_0$ (curves 2–5), the potential energy plot has a maximum at the $\theta = 0$ point, which corresponds to the unstable position.

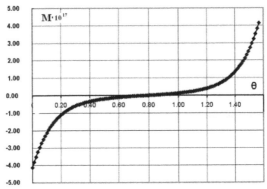

FIGURE 12.16   The dependence of the moment M[Nm] of the interaction force between cubic nanoparticles 1–3 (see in Fig. 12.9) on the angle of their relative rotation θ [rad].

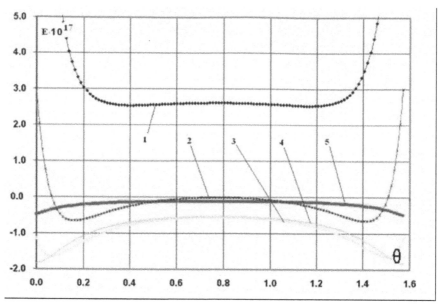

FIGURE 12.17   The plots of the change of the potential energy E[Nm] for the interaction of two cubic nanoparticles depending on the angle of their relative rotation θ [rad] and the distance between them (positions of the nanoparticles 1–3, Fig. 12.9).

The carried out theoretical analysis is confirmed by the works of the scientists from New Jersey University and California University in Berkeley who experimentally found the self-organization of the cubic microparticles of plumbum zirconate-titanate (PZT) [28]: the ordered groups of cubic microcrystals from PZT obtained by hydrothermal synthesis formed a flat layer of particles on the air-water interface, where the particle occupied the more stable position corresponding to position 2, 3 in Fig. 12.15.

Thus, the analysis of the interaction of two cubic nanoparticles has shown that different variants of their final stationary state of equilibrium are possible, in which the principal vectors of forces and moments are zero. However, there are both stable and unstable stationary states of this system: nanoparticle positions 2, 3 are stable, and positions 3, 4 and 2–5 have limited stability or they are unstable depending on the distance between the nanoparticles.

Note that for the structures consisting of a large number of nanoparticles, there can be a quantity of stable stationary and unstable forms of equilibrium. Accordingly, the stable and unstable nanostructures of composite materials can appear. The search and analysis of the parameters determining the formation of stable nano systems is an urgent task.

It is necessary to note, that the method offered has restrictions. This is explained by change of the nanoparticles form and accordingly variation of interaction pair potential during nanoparticles coming together at certain conditions.

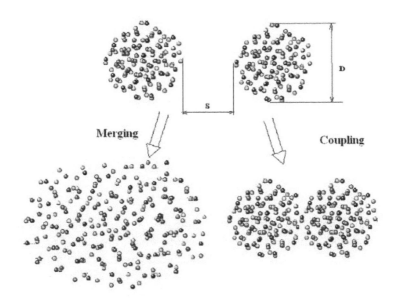

FIGURE 12.18 Different type of nanoparticles connection (merging and coupling).

The merge (accretion [4]) of two or several nanoparticles into a single whole is possible (Fig. 12.18). Change of a kind of connection cooperating nanoparticles (merging or coupling in larger particles) depending on the sizes, it is possible to explain on the basis of the analysis of the energy change graph of connection nanoparticles (Fig. 12.19). From (Fig. 12.19) follows, that, though with the size increasing of a particle energy of nanoparticles connection E_{np} grows also, its size in comparison with superficial energy E_s of a particle sharply increases at reduction of the sizes nanoparticles. Hence, for finer particles energy of connection can appear sufficient for destruction of their configuration under action of a mutual attraction and merging in larger particle.

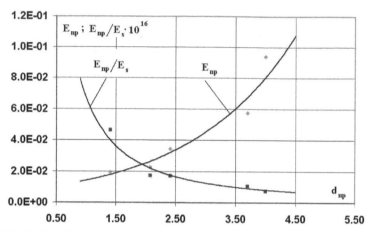

FIGURE 12.19   Change of energy of nanoparticles connection E_{np} [Nm] and E_{np} ration to superficial energy E_s depending on nanoparticles diameter d [nm]. Points designate the calculated values. Continuous lines are approximations.

Spatial distribution of particles influences on rate of the forces holding nanostructures, formed from several nanoparticles, also. On Fig. 12.20, the chain nanoparticles, formed is resulted at coupling of three nanoparticles (from 512 atoms everyone), located in the initial moment on one line. Calculations have shown that in this case nanoparticles form a stable chain. Thus, particles practically do not change the form and cooperate on "small platforms."

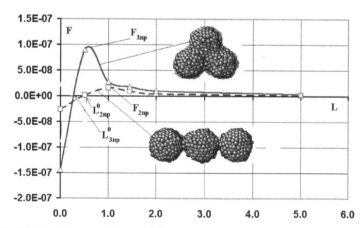

FIGURE 12.20   Change of force F [N] of 3 nanoparticles interaction, consisting of 512 atoms everyone, and connected among themselves on a line and on the beams missing under a corner of 120 degrees, accordingly, depending on distance between them L [nm].

In the same figure the result of connection of three nanoparticles, located in the initial moment on a circle and consisting of 256 atoms everyone is submitted. In this case particles incorporate among themselves "densely," contacting on a significant part of the external surface.

Distance between particles at which they are in balance it is much less for the particles collected in group ($L_{3np}^0 < L_{2np}^0$). It confirms also the graph of forces from which it is visible, that the maximal force of an attraction between particles in this case (is designated by a continuous line) in some times more, than at an arrangement of particles in a chain (dashed line) $F_{3np} > F_{2np}$.

Experimental investigation of the spatial structures formed by nanoparticles [4], confirm that nanoparticles gather to compact objects. Thus the internal nuclear structure of the connections area of nanoparticles considerably differs from structure of a free nanoparticle.

Nanoelements kind of interaction depends strongly on the temperature. Figure 12.21 shows the picture of the interaction of nanoparticles at different temperatures (Fig. 12.22). It is seen that with increasing temperature the interaction of changes in sequence: coupling (1, 2), merging (3, 4). With further increase in temperature the nanoparticles dispersed.

FIGURE 12.21 Change of nanoparticles connection at increase in temperature.

In the conclusion to the section we will consider problems of dynamics of nanoparticles. The analysis of interaction of nanoparticles among themselves also allows drawing a conclusion on an essential role in this process of energy of initial movement of particles. Various processes at interaction of the nanoparticles, moving with different speed, are observed: the processes of agglomerate formation, formation of larger particles at merge of the smaller size particles, absorption by large particles of the smaller ones, dispersion of particles on separate smaller ones or atoms.

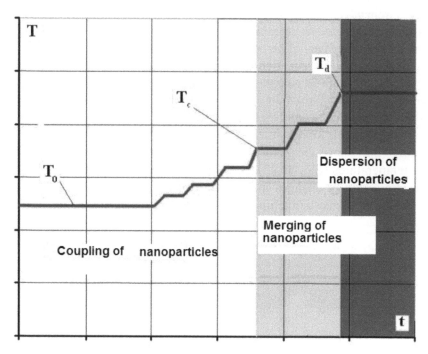

FIGURE 12.22   Curve of temperature change.

For example, in Fig. 12.23 the interactions of two particles are moving towards each other with different speed are shown. At small speed of moving is formed steady agglomerate (Fig. 12.23).

In a Fig. 12.23(left) is submitted interaction of two particles moving towards each other with the large speed. It is visible, that steady formation in this case is not appearing and the particles collapse.

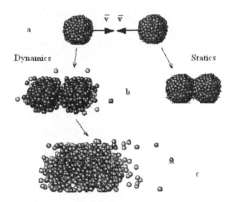

FIGURE 12.23 Pictures of dynamic interaction of two nanoparticles: (a) an initial configuration nanoparticles; (b) nanoparticles at dynamic interaction; (c) the "cloud" of atoms formed because of dynamic destruction two nanoparticles.

Nature of interaction of nanoparticles, along with speed of their movement, essentially depends on a ratio of their sizes. In Fig. 12.24, pictures of interaction of two nanoparticles of zinc of the different size are presented. As well as in the previous case of a nanoparticle move from initial situation (*1*) towards each other. At small initial speed, nanoparticles incorporate at contact and form a steady conglomerate (*2*).

FIGURE 12.24 Pictures of interaction of two nanoparticles of zinc: *1* – initial configuration of nanoparticles; *2* – connection of nanoparticles, *3,4* absorption by a large nanoparticle of a particle of the smaller size, *5* – destruction of nanoparticles at blow.

At increase in speed of movement larger nanoparticle absorbs smaller, and the uniform nanoparticle (*3, 4*) is formed. At further increase in speed of movement of nanoparticles, owing to blow, the smaller particle intensively takes root in big and destroys its.

The given examples show that use of dynamic processes of pressing for formation of nano composites demands a right choice of a mode of the loading providing integrity of nanoparticles. At big energy of dynamic loading, instead of a nanocomposite with dispersion corresponding to the initial size of nanoparticles, the nanocomposite with much larger grain that will essentially change properties of a composite can be obtained.

12.6   CONCLUSIONS

In conclusion, the following basic regularities of the nanoparticle formation and self-organization should be noted.

1. The existence of several types of the forms and structures of nanoparticles is possible depending on the thermodynamic conditions.
2. The absence of the crystal nucleus in small nanoparticles.
3. The formation of a single (ideal) crystal nucleus defects on the nucleus surface connected to the amorphous shell.
4. The formation of the poly crystal nucleus with defects distributed among the crystal grains with low atomic density and change inter atomic distances. In addition, the grain boundaries are nonequilibrium and contain a great number of grain-boundary defects.
5. When there is an increase in sizes, the structure of nanoparticles is changing from amorphous to roentgen-amorphous and then into the crystalline structure.
6. The formation of the defect structures of different types on the boundaries and the surface of a nanoparticle.
7. The nanoparticle transition from the globe-shaped to the crystal-like shape.
8. The formation of the "regular" and "irregular" shapes of nanoparticles depending on the number of atoms forming the nanoparticle and the relaxation conditions (the rate of cooling, first of all).
9. The structure of a nanoparticle is strained because of the different distances between the atoms inside the nanoparticle and in its surface layers.
10. The systems of nanoparticles can to form stable and unstable nanostructures.

ACKNOWLEDGMENTS

This work was carried out with financial support from the Research Program of the Ural Branch of the Russian Academy of Sciences: the projects 12-P-12-2010, 12-C-1-1004, 12-T-1-1009 and was supported by the grants of Russian Foundation for Basic Research (RFFI) 04-01-96017-r2004ural_a; 05-08-50090-a; 07-01-96015-r_ural_a; 08-08-12082-ofi; 11-03-00571-a.

The author is grateful to his young colleagues Dr. A.A. Vakhrushev and Dr. A.Yu. Fedotov for active participation in the development of the software complex, calculations and analyzing of numerous calculation results.

The calculations were performed at the Joint Supercomputer Center of the Russian Academy of Sciences.

KEYWORDS

- **Amorphous**
- **Atomistic Simulation**
- **Nano Particles**
- **Nanostructural elements**
- **Roentgen-amorphous**
- **The crystal nucleus**

REFERENCES

1. Qing-Qing, Ni, Yaqin, Fu, & Masaharu Iwamoto (2004). "Evaluation of Elastic Modulus of Nano Particles in PMMA/Silica Nano Composites," *Journal of the Society of Materials Science*, Japan, *53(9)*, 956–961.
2. Ruoff, R. S., & Nicola Pugno, M. (2004). "*Strength of Nanostructures*," Mechanics of the twenty-first century Proceeding of the 21th International Congress of Theoretical and Applied Mechanics, Warsaw **Springer,** 303–311.
3. Diao, J., Gall, K., & Dunn, M. L. (2004). "Atomistic Simulation of the Structure and Elastic Properties of Gold Nano Wires," *Journal of the Mechanics and Physics of Solids, 52(9),* 1935–1962.
4. Dingreville, R., Qu, J., & Cherkaoui, M. (2004). "Surface Free Energy and its Effect on the Elastic Behavior of Nano-Sized Particles, Wires and Films," *Journal of the Mechanics and Physics of Solids, 53(8)*, 1827–1854.
5. Duan, H. L., Wang, J., Huang, Z. P., & Karihaloo, B. L. (2005). "Size-Dependent Effective Elastic Constants of Solids Containing Nano in homogeneities with Interface Stress," *Journal of the Mechanics and Physics of Solids, 53(7),* 1574–1596.
6. Gusev, A. I., & Rempel, A. A. (2001). "*Nanocrystalline Materials*" Moscow Physical Mathematical literature, (in Russian).

7. Hoare, M. R. (1987). "Structure and Dynamics of Simple Micro Clusters" *Ach Chem Phys, 40,* 49–135.

8. Brooks, B. R., Bruccoleri, R. E., Olafson, B. D., States, D. J., Swaminathan, S., & Karplus, M. (1983). "Charmm a Program for Macromolecular Energy Minimization and Dynamics Calculations," J Comput Chemistry, 4(2), 187–217.

9. Friedlander, S. K. (1999). "Polymer like Behavior of Inorganic Nanoparticle Chain Aggregates," *Journal of Nanoparticle Research, 1,* 9–15.

10. Grzegorczyk, M., Rybaczuk, M. & Maruszewski, K. (2004). "Ballistic Aggregation an Alternative Approach to Modeling of Silica Sol–Gel Structures" Chaos Solitons and Fractals, 19, 1003–1011.

11. Shevchenko, E. V., Talapin, D. V., Kotov, N. A., Brien, S. O., & Murray, C. B. (2006). "Structural Diversity in Binary Nanoparticle Super Lattices" Nature Letter, 439, 55–59.

12. Kang, Z. C., & Wang, Z. L. (1996). "On Accretion of Nano Size Carbon Spheres," *Journal Physical Chemistry, 100,* 5163–5165.

13. Melikhov, I. V., & Bozhevol'nov, V. E. (2003). "Variability and Self-Organization in Nano Systems," *Journal of Nanoparticle Research, 5,* 465–472

14. Kim, D., & Lu, W. (2004). "Self-Organized Nanostructures in Multi-Phase Epilayers" *Nanotechnology, 15,* 667–674.

15. Kurt Geckeler E. (2005). "Novel Super Molecular Nanomaterials from Design to Reality," Proceeding of the 12 Annual International Conference on Composites, Nano Engineering, Tenerife, Spain, August 1–6, CD Rom Edition.

16. Vakhrouchev, A. V., & Lipanov A. M. (1992). "A Numerical Analysis of the Rupture of Powder Materials under the Power Impact Influence," *Computer & Structures, 1/2 (44),* 481–486.

17. Vakhrouchev, A. V. (2004). "Modeling of Static and Dynamic Processes of Nanoparticles Interaction," CD-ROM Proceeding of the 21th International Congress of Theoretical and Applied Mechanics, ID12054, Warsaw, Poland.

18. Vakhrouchev, A. V. (2005). "Simulation of Nanoparticles Interaction," Proceeding of the 12 Annual International Conference on Composites, Nano Engineering, Tenerife, Spain, August 1–6, CD Rom Edition.

19. Vakhrouchev, A. V. (2006). "Simulation of Nano-elements Interactions and Self-assembling," *Modeling and Simulation in Materials Science and Engineering, 14,* 975–991

20. Vakhrouchev, A. V., & Lipanov, A. M. (2007). Numerical Analysis of the Atomic structure and Shape of Metal Nanoparticles, *Computational Mathematics and Mathematical Physics, 47(10),* 1702–1711.

21. Vakhrouchev, A. V. (2008). Modeling of the Process of Formation and use of Powder Nano Composites, Composites with Micro and Nano-Structures Computational Modeling and Experiments, Computational Methods in Applied Sciences Series, Barcelona, Spain, Springer Science, 9, 107–136.

22. Vakhrouchev A. V. (2009). Modeling of the Nano Systems formation by the Molecular Dynamics, Meso Dynamics and Continuum Mechanics Methods, Multidiscipline Modeling in Material and Structures, 5(2), 99–118.

23. Vakhrouchev, A. V. (2008). Theoretical Bases of Nano Technology Application to Thermal Engines and Equipment, Izhevsk, Institute of Applied Mechanics, Ural Branch of the Russian Academy of Sciences, 212p (in Russian).

24. Alikin, V. N., Vakhrouchev, A. V., Golubchikov, V. B., Lipanov, A. M., & Serebrennikov, S. Y. (2010). Development and Investigation of the Aerosol Nanotechnology, Moscow Mashinostroenie, 196p, (in Russian).

25. Heerman, W. D. (1986). "Computer Simulation Methods in Theoretical Physics" Berlin Springer-Verlag.
26. Verlet, L (1967). "Computer "Experiments" on Classical Fluids I. Thermo Dynamical Properties of Lennard-Jones molecules," *Phys. Rev., 159*, 98–103.
27. Korn, G. A., & Korn, M. T. (1968). "Mathematical Handbook," New York: McGraw-Hill Book Company.
28. Self-organizing of microparticles piezoelectric materials, News of Chemistry, datenews. php.htm

A DETAILED REVIEW ON STRUCTURE AND PROPERTIES OF HIGH PERFORMANCE CARBON NANOTUBE/POLYMER COMPOSITES

A. K. HAGHI and G. E. ZAIKOV

CONTENTS

ABSTRACT

In this chapter, new trends in computational chemistry and computational mechanics for the prediction of the structure and properties of CNT materials are presented simultaneously.

13.1   INTRODUCTION

It has been known that the mechanical properties of polymeric materials like stiffness and strength can be engineered by producing composites that are composed of different volume fraction of one or more reinforcing phases. In traditional form, polymeric materials have been reinforced with carbon, glass, basalt, ceramic and aramid microfibers to improve their mechanical properties. These composite materials have been used in many applications in automotive, aerospace and mass transit. As time has proceeded, a practical accomplishment of such composites has begun to change from micro scale composites to nanocomposite, taking advantages of better mechanical properties. While some credit can be attributed to the intrinsic properties of the fillers, most of these advantages stem from the extreme reduction in filler size combined with the large enhancement in the specific surface area and interfacial area they present to the matrix phase. In addition, since traditional composites use over 40wt% of the reinforcing phase, the dispersion of just a few weight percentages of nano fillers into polymeric matrices could lead to dramatic changes in their mechanical properties. One of the earliest nano filler that witch have received significant and shown super mechanical properties is Carbon nanotube (CNT), because of their unique properties, CNTs have a wide range of potentials for engineering applications due to their exceptional mechanical, physical, electrical properties and geometrical characteristics consisting of small diameter and high aspect ratio. It has shown that dispersion of a few weight percentages of nanotubes in a matrix dramatically increase mechanical, thermal and electrical properties of composite materials. Development of CNT-nano composites requires a good understanding of CNT's and CNT's nanocomposite properties. Because of the huge cost and technological difficulties associated with experimental analysis at the scale of nano, researchers are encouraged to employ computational methods for simulating the behavior of nanostructures like CNTs from different mechanical points of view.

To promote the design and development of CNT-nanocomposite materials, structure and property relationships must be recognized that predict the bulk mechanical of these materials as a function of the molecular and micro structure mechanical properties of nano structured materials can be calculated by a select set of computational methods. These modeling methods extend across a wide range of length and time scales, as shown in Fig. 13.1. For the smallest length and time scales, a complete understanding of the behavior of materials requires theoretical and computational tools that span the atomic-scale detail of first principles methods (density functional theory, molecular dynamics, and Monte Carlo methods). For the largest length and time scales, computational mechanics is used to predict the mechanical behavior of materials and engineering structures. And the coarser grained description provided by continuum equations. However, the intermediate length and time scales do not have general modeling methods that are as well developed as those on the smallest and largest time and length scales. Therefore, recent efforts have focused on combining traditional methodologies and continuum descriptions within a unified multi scale framework. Multi scale modeling techniques are employed, which take advantage of computational chemistry and computational mechanics methods simultaneously for the prediction of the structure and properties of materials.

As illustrated in Fig. 13.1, in each modeling methods have extended classes of related modeling tools that are shown in a short view in a diagram in Fig. 13.2.

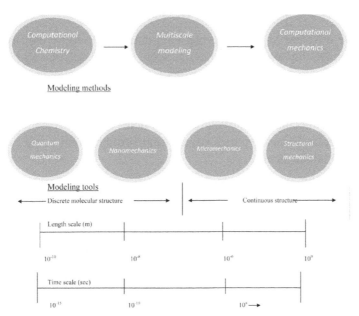

FIGURE 13.1 Different length and time scale used in determination mechanical properties of polymer nano-composite.

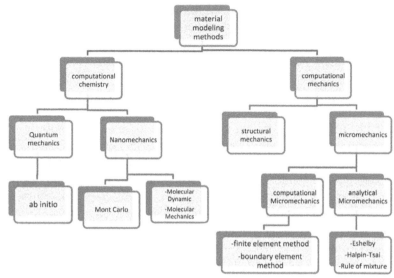

FIGURE 13.2   Diagram of material modeling methods.

13.2   MATERIAL MODELING METHODS

There are different of modeling methods currently used by the researches. They plan not only to simulate material behavior at a particular scale of interest but also to assist in developing new materials with highly desirable properties. These scales can range from the basic atomistic to the much coarser continuum level. The hierarchy of the modeling methods consists quantum mechanics, molecular dynamic, micromechanics and finally continuum mechanics, that could be categorized in tow main group: atomistic modeling and computational mechanics.

13.2.1   COMPUTATIONAL CHEMISTRY (ATOMISTIC MODELING)

The atomistic methods usually employ atoms, molecules or their group and can be classified into three main categories, namely the quantum mechanics (QM), molecular dynamics (MD) and Monte Carlo (MC). Other atomistic modeling techniques such as tight bonding molecular dynamics (TBMD), local density (LD), dissipative particle dynamics (DPD), lattice Boltzmann (LB), Brownian dynamics (BD), time-dependent Ginzburg–Lanau method,

Morse potential function model, and modified Morse potential function model were also applied afterwards.

13.2.1.1   QUANTUM MECHANICS

The observable properties of solid materials are governed by quantum mechanics, as expressed by solutions of a Schrödinger equation for the motion of the electrons and the nuclei. However, because of the inherent difficulty of obtaining even coarsely approximate solutions of the full many body Schrödinger equation, one typically focuses on reduced descriptions that are believed to capture the essential energetic of the problem of interest. Two main quantum mechanics method are "AB initio" and "density function method (DFT)."

Unlike most materials simulation methods that are based on classical potentials, the main advantages of AB initio methods, which is based on first principles density functional theory (without any adjustable parameters), are the generality, reliability, and accuracy of these methods. They involve the solution of Schrödinger's equation for each electron, in the self-consistent potential created by the other electrons and the nuclei. AB initio methods can be applied to a wide range of systems and properties. However, these techniques are computationally exhaustive, making them difficult for simulations involving large numbers of atoms. There are three widely used procedures in AB initio simulation. These procedures are single point calculations, geometry optimization, and frequency calculation. Single point calculations involve the determination of energy and wave functions for a given geometry. This is often used as a preliminary step in a more detailed simulation. Geometry calculations are used to determine energy and wave functions for an initial geometry, and subsequent geometries with lower energy levels. A number of procedures exist for establishing geometries at each calculation step. Frequency calculations are used to predict Infrared and Raman intensities of a molecular system. AB initio simulations are restricted to small numbers of atoms because of the intense computational resources that are required.

AB initio techniques have been used on a limited basis for the prediction of mechanical properties of polymer based nano structured composites.

13.2.1.2   MOLECULAR DYNAMICS

Molecular Dynamic is a simulation technique that used to estimate the time depended physical properties of a system of interacting particles (e.g., atoms, molecules, etc.) by predict the time evolution. MD simulation used to investigating the structure, dynamics, and thermodynamics of individual molecules.

There are two basic assumptions made in standard molecular dynamics simulations. Molecules or atoms are described as a system of interacting material points, whose motion is described dynamically with a vector of instantaneous positions and velocities. The atomic interaction has a strong dependence on the spatial orientation and distances between separate atoms. This model is often referred to as the soft sphere model, where the softness is analogous to the electron clouds of atoms.

- No mass changes in the system. Equivalently, the number of atoms in the system remains the same.

The atomic positions velocities and accelerations of individual particles that vary with time. The described by track that MD simulation generates and then used to obtain average value of system such as energy, pressure and temperature.

The main three parts of MD simulation are:

- initial conditions (e.g., initial positions and velocities of all particles in the system);
- The interaction potentials between particles to represent the forces among all the particles;
- The evolution of the system in time by solving of classical Newtonian equations of motion for all particles in the system.

The equation of motion is generally given by Eq. (1).

$$\vec{F}_i(t) = m_i \frac{d^2 \vec{r}_i}{dt^2} \tag{1}$$

where is the force acting on the i-th atom or particle at time t and is obtained as the negative gradient of the interaction potential U, m_i is the atomic mass and the atomic position. The interaction potentials together with their parameters, describe how the particles in a system interact with each other (so-called force field). Force field may be obtained by quantum method (e.g., AB initio), empirical method (e.g., Lennard-Jones, Mores, and Born-Mayer) or quantum-empirical method (e.g., embedded atom model, glue model, bond order potential).

The criteria for selecting a force field include the accuracy, transferability and computational speed. A typical interaction potential U may consist of a number of bonded and nonbonded interaction terms:

$$U(\overrightarrow{r_1}, \overrightarrow{r_2}, \overrightarrow{r_3}, \ldots, \overrightarrow{r_n})$$

$$= \sum_{i_{bond}}^{N_{bond}} U_{bond}(i_{bond}, \overrightarrow{r_a}, \overrightarrow{r_b}) +$$

$$\sum_{i_{angle}}^{N_{angle}} U_{angle}(i_{angle}, \overrightarrow{r_a}, \overrightarrow{r_b}, \overrightarrow{r_c}) + \sum_{i_{torsion}}^{N_{torsion}} U_{torsion}(i_{torsion}, \overrightarrow{r_a}, \overrightarrow{r_b}, \overrightarrow{r_c}, \overrightarrow{r_d})$$

$$+ \sum_{i_{inversion}}^{N_{inversion}} U_{inversion}(i_{inversion}, \overrightarrow{r_a}, \overrightarrow{r_b}, \overrightarrow{r_c}, \overrightarrow{r_d}) \qquad (2)$$

$$+ \sum_{i=1}^{N-1} \sum_{j>i}^{N} U_{vdw}(i, j, \overrightarrow{r_a}, \overrightarrow{r_b})$$

$$+ \sum_{i=1}^{N-1} \sum_{j>i}^{N} U_{electrostatic}(i, j, \overrightarrow{r_a}, \overrightarrow{r_b})$$

The first four terms represent bonded interactions, that is, bond stretching U_{bond}, bond-angle bend U_{angle}, dihedral angle torsion $U_{torsion}$ and inversion interaction $U_{inversion}$. Vander Waals energy U_{vdw} and electrostatic energy $U_{electrostatic}$ are nonbonded interactions. In the equation, are the positions of the atoms or particles specifically involved in a given interaction;,, and illustrate the total numbers of interactions in the simulated system; ,, and presented an individual interaction each of them. There are many algorithms like *varlet*, *velocity varlet*, *leap-frog* and *Beeman*, for integrating the equation of motion, all of them using finite difference methods, and assume that the atomic position , velocities and accelerations can be approximated by a *Taylor series expansion*:

$$\overrightarrow{r}(t + \delta t) = \overrightarrow{r}(t) + \overrightarrow{v}(t)\delta t + \frac{1}{2}\overrightarrow{a}(t)\delta^2 t + \cdots \qquad (3)$$

$$\overrightarrow{v}(t + \delta t) = \overrightarrow{v}(t) + \overrightarrow{a}(t)\delta t + \frac{1}{2}\overrightarrow{b}(t)\delta^2 t + \cdots \qquad (4)$$

$$\overrightarrow{a}(t + \delta t) = \overrightarrow{a}(t) + \overrightarrow{b}(t)\delta t + \cdots \qquad (5)$$

The *varlet algorithm* is probably the most widely used method. It uses the positions and accelerations at time *t*, and the positions from the previous step *(t–δ)* to calculate the new positions at *(t + δt)*, so:

$$\overrightarrow{r}(t + \delta t) = \overrightarrow{r}(t) + \overrightarrow{v}(t)\delta t + \frac{1}{2}\overrightarrow{a}(t)\delta t^2 + \cdots \qquad (6)$$

$$\vec{r}(t - \delta t) = \vec{r}(t) - \vec{v}(t)\delta t + \frac{1}{2}\vec{a}(t)\delta t^2 + \cdots \tag{7}$$

$$\vec{r}(t + \delta t) = 2\vec{r}(t) - \vec{r}(t - \delta t) + \vec{a}(t)\delta t^2 + \cdots \tag{8}$$

The velocities at time t and can be respectively estimated.

$$\vec{v}(t) = [\vec{r}(t + \delta t) - \vec{r}(t - \delta t)]/2\delta t \tag{9}$$

$$\vec{v}(t + 1/2\delta t) = [\vec{r}(t + \delta t) - \vec{r}(t - \delta t)]/\delta t \tag{10}$$

The advantages of this method are *time-reversible* and *good energy conservation* properties, where disadvantage is *low memory storage* because the velocities are not included in the time integration. However, removing the velocity from the integration introduces numerical in accuracies method, namely the *velocity Verlet* and *Verlet leap-frog algorithms* as mentioned before, which clearly involve the velocity in the time evolution of the atomic coordinates.

13.2.1.3   MONTE CARLO

Monte Carlo technique (*Metropolis method*) use random number from a given probability distribution to generate a sample population of the system from which one can calculate the properties of interest a MC simulation usually consists of three typical steps:
- The physical problem is translated into an analogous probabilistic or statistical model.
- The probabilistic model is solved by a numerical stochastic sampling experiment.
- The obtained data are analyzed by using statistical methods.

Athwart MD which provides information for nonequilibrium as well as equilibrium properties, MC gives only the information on equilibrium properties (e.g., free energy, phase equilibrium). In a NVT ensemble with N atoms, new formation the change in the system *Hamiltonian* (ΔH) by randomly or systematically moving one atom from position $i \to j$ can calculated.

$$\Delta H = H(j)-H(i) \tag{11}$$

where $H(i)$ and $H(j)$ are the *Hamiltonian* associated with the original and new configuration, respectively.

The $\Delta H < 0$ shows the state of lower energy for system. So, the movement is accepted and the new position is stable place for atom.

For $\Delta H \geq 0$ the move to new position is accepted only with a certain probability P which is given by

$$Pi \rightarrow j \propto exp\left(-\frac{\Delta H}{K_B T}\right) \tag{12}$$

is the Boltzmann constant.

According to *Metropolis* et al. a random number ζ between 0 and 1 could be generated and determine the new configuration according to the following rule:

$$\text{For } \zeta \leq exp\left(-\frac{\Delta H}{K_B T}\right) \zeta; \text{ the move is accepted;} \tag{13}$$

$$\text{For } \zeta > exp\left(-\frac{\Delta H}{K_B T}\right); \text{ the move is not accepted.} \tag{14}$$

If the new configuration is rejected, repeats the process by using other random chosen atoms.

In a μVT ensemble, a new configuration j by arbitrarily chosen and it can be exchanged by an atom of a different kinds. This method affects the chemical composition of the system and the move is accepted with a certain probability. However, the energy, ΔU, will be changed by change in composition.

If $\Delta U < 0$, the move of compositional changes is accepted. However, if $\Delta U \geq 0$, the move is accepted with a certain probability which is given by:

$$Pi \rightarrow j \propto exp\left(-\frac{\Delta U}{K_B T}\right) \tag{15}$$

where ΔU is the change in the sum of the mixing energy and the chemical potential of the mixture. If the new configuration is rejected one counts the original configuration as a new one and repeats the process by using some other arbitrarily or systematically chosen atoms. In polymer nano composites, MC methods have been used to investigate the molecular structure at nanoparticle surface and evaluate the effects of various factors.

13.2.2 COMPUTATIONAL MECHANICS

The continuum material that assumes is continuously distributed throughout its volume have an average density and can be subjected to body forces such as gravity and surface forces. Observed macroscopic behavior is usually illustrated by ignoring the atomic and molecular structure. The basic laws for continuum model are:

- continuity, drawn from the conservation of mass;
- equilibrium, drawn from momentum considerations and Newton's second law;
- the moment of momentum principle, based on the model that the time rate of change of angular momentum with respect to an arbitrary point is equal to the resultant moment;
- conservation of energy, based on the first law of thermodynamics;
- conservation of entropy, based on the second law of thermodynamics.

These laws provide the basis for the continuum model and must be coupled with the appropriate constitutive equations and the equations of state to provide all the equations necessary for solving a continuum problem. The continuum method relates the deformation of a continuous medium to the external forces acting on the medium and the resulting internal stress and strain. Computational approaches range from simple closed-form analytical expressions to micromechanics and complex structural mechanics calculations based on beam and shell theory. In this section, we introduce some continuum methods that have been used in polymer nano composites, including micromechanics models (e.g., Halpin-Tsai model, Mori-Tanaka model and finite element analysis.) and the semicontinuum methods like equivalent-continuum model and will be discussed in the next part.

13.2.2.1 MICROMECHANICS

Micromechanics are a study of mechanical properties of unidirectional composites in terms of those of constituent materials. In particular, the properties to be discussed are elastic modulus, hydrothermal expansion coefficients and strengths. In discussing composites properties it is important to define a *volume element,* which is small enough to show the microscopic structural details, yet large enough to present the overall behavior of the composite. Such a volume element is called the *Representative Volume Element (RVE).* A simple representative volume element can consists of a fiber embedded in a matrix block, as shown in Fig. 13.3.

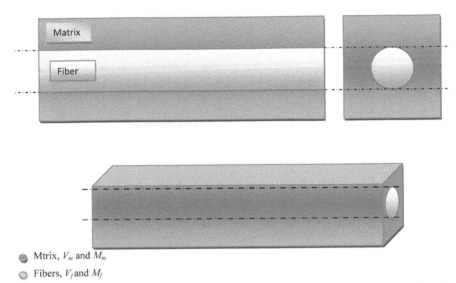

● Mtrix, V_m and M_m

○ Fibers, V_f and M_f

FIGURE 13.3   A Representative Volume Element. The total volume and mass of each constituent are denoted by V and M, respectively. The subscripts m and f stand for matrix and fiber, respectively.

One a representative volume element is chosen, proper boundary conditions are prescribed. Ideally, these boundary conditions must represent the in situ state of stress and strain within the composite. That is, the prescribed boundary conditions must be the same as those if the representative volume element were actually in the composite.

Finally, a prediction of composite properties follows from the solution of the foregoing boundary value problem. Although the procedure involved is conceptually simple, the actual solution is rather difficult. Consequently, many assumption and approximation have been introduces, and therefore various solution are available.

13.2.2.1.1   BASIC CONCEPTS

Micromechanics models usually used to reinforced polymer composites, based on follow basic assumptions:

1.   linear elasticity of fillers and polymer matrix;

2. the reinforcement are axis-symmetric, identical in shape and size, and can be characterized by parameters such as length-to-diameter ratio (aspect ratio);

3. Perfect bonding between reinforcement and polymer interface and the ignorance of interfacial slip, reinforcement and polymer de bonding or matrix cracking.

Consider a composite of mass M and volume V, illustrated schematically in Fig. 13.3, V is the volume of a RVE, since the composite is made of fibers and matrix, the mass M is the sum of the total mass M_f of fibers and mass M_m of matrix:

$$M = M_f + M_m \tag{16}$$

Equation (16) is valid regardless of voids, which may be present. However, the composite volume V includes the volume V_v of voids so that:

$$V = V_f + V_m + V_v \tag{17}$$

Dividing Eqs. (16) and (17), leads to the following relation for the mass fraction and volume fractions:

$$m_f + m_m = 1 \tag{18}$$

$$v_f + v_m + v_v = 1 \tag{19}$$

The composite density ρ calculated as follows:

$$\rho = \frac{M}{V} = \frac{(\rho_f v_f + \rho_m v_m)}{V} = \rho_f v_f + \rho_m v_m \tag{20}$$

$$\rho = \frac{1}{m_f \big/ \rho_f + m_m \big/ \rho_m + v_v \big/ \rho} \tag{21}$$

These equations can be used to determine the void fraction:

$$v_v = 1 - \rho \left(\frac{m_f}{\rho_f} + \frac{m_m}{\rho_m} \right) \tag{22}$$

The mass fraction of fibers can be measured by removing the matrix. Based on the first concept, the linear elasticity, the linear relationship between the total stress and infinitesimal strain tensors for the reinforcement and matrix as expressed by the following constitutive equations:

$$\sigma_f = C_f \varepsilon_f \tag{23}$$

$$\sigma_m = C_m \varepsilon_m \tag{24}$$

where C is the stiffness tensor.

The second concept is the average stress and strain. While the stress field and the corresponding strain field are usually nonuniform in polymer composites, the average stress and strain are then defined over the representative averaging volume V, respectively. Hypothesize the stress field in the RVE is, Then composite stress and is defined by:

$$\bar{\sigma} = \frac{1}{V}\int \sigma_i dv = \frac{1}{V}[\int_{v_f} \sigma_i dv + \int_{v_m} \sigma_i dv + \int_{v_v} \sigma_i dv] \tag{25}$$

$$\bar{\sigma}_{fi} = \frac{1}{V_f}\int_{v_f} \sigma_i dV \ , \ \bar{\sigma}_{mi} = \frac{1}{V_m}\int_{v_m} \sigma_i dV \tag{26}$$

Because of no stress is transmitted in the voids, in and so:

$$\bar{\sigma}_i = v_f \bar{\sigma}_{fi} + v_m \bar{\sigma}_{mi} \tag{27}$$

Where is composite average stress, is fibers average stress and is matrix average stress. Similarly to the composite stress, the composite strain is defined as the volume average strain, and is obtained as:

$$\bar{\varepsilon}_i = v_f \bar{\varepsilon}_{fi} + v_m \bar{\varepsilon}_{mi} + v_v \bar{\varepsilon}_{vi} \tag{28}$$

Despite the stress, the void in strain does not vanished; it is defined in term of the boundary displacements of the voids. So, because of the void fraction is usually negligible. Therefore, last term in Eq. (26) could be neglected and the equation corrected to:

$$\bar{\varepsilon}_i = v_f \bar{\varepsilon}_{fi} + v_m \bar{\varepsilon}_{mi} \tag{29}$$

The average *stiffness* of the composite is the tensor C that related the average strain to the average stress as follow equation:

$$\bar{\sigma} = C\,\bar{\varepsilon}$$
(30)

The average *compliance* S is defined in this way:

$$\bar{\varepsilon} = S\,\bar{\sigma}$$
(31)

Another important concept is the *strain concentration* and *stress concentration* tensors A and B, which are basically the ratios between the average reinforcement strain or stress and the corresponding average of the composites.

$$\bar{\varepsilon}_f = A\bar{\varepsilon}$$
(32)

$$\bar{\sigma}_f = B\bar{\sigma}$$
(33)

Finally, the average composite stiffness can be calculated from the strain concentration tensor A and the reinforcement and matrix properties:

$$C = C_m + v_f(C_f - C_m)A$$
(34)

13.2.2.1.2   HALPIN–TSAI MODEL

Halpin-Tsai theory is used for prediction elastic modulus of unidirectional composites as function of aspect ratio. The longitudinal stiffness, E_{11} and transverse modulus, E_{22}, are expressed in the following general form:

$$\frac{E}{E_m} = \frac{1 + \zeta \eta v_f}{1 - \eta v_f}$$
(35)

where E and E_m are modulus of composite and matrix respectively, is fiber volume fraction and η is given by this equation:

$$\eta = \frac{\dfrac{E}{E_m} - 1}{\dfrac{E_f}{E_m} + \zeta_f} \tag{36}$$

where fiber modulus is and ζ is shape parameter that depended on reinforcement geometry and loading direction. For E_{11} calculation, ζ is equal to L/t where L is length and t is thickness of reinforcement, for E_{22}, ζ is equal to w/t where w is width of reinforcement.

For →0, the Halpin-Tsai theory converged to inversed *rule of mixture* for stiffness.

$$\frac{1}{E} = \frac{v_f}{E_f} + \frac{1 - v_f}{E_m} \tag{37}$$

If →∞, the Halpin-Tsai converge to rule of mixture.

$$E = E_f v_f + E_m (1 - v_f) \tag{38}$$

13.2.2.1.3   MORI-TANAKA MODEL

The Mori-Tanaka model is uses for prediction an elastic stress field for in and around an ellipsoidal reinforcement in an infinite matrix. This method is based on Eshebly's model. Longitudinal and transverse elastic modulus, E_{11} and E_{22}, for isotropic matrix and directed spherical reinforcement are:

$$\frac{E_{11}}{E_m} = \frac{A_0}{A_0 + v_f (A_1 + 2v_0 A_2)} \tag{39}$$

$$\frac{E_{22}}{E_m} = \frac{2A_0}{2A_0 + v_f (-2A_3 + (1 - v_0 A_4) + (1 + v_0) A_5 A_0)} \tag{40}$$

where is the modulus of the matrix, is the volume fraction of reinforcement, is the Poisson's ratio of the matrix, parameters, A_0, A_1,...,A_5 are functions of the Eshelby's tensor.

13.2.2.2   FINITE ELEMENT METHODS

The traditional framework in mechanics has always been the continuum. Under this framework, materials are assumed to be composed of a divisible continuous medium, with a constitutive relation that remains the same for a wide range of system sizes. Continuum equations are typically in the form of deterministic or stochastic partial differential equation (FDE's). The underling atomic structure of matter is neglected altogether and is replaced with a continuous and differentiable mass density. Similar replacement is made for other physical quantities such as energy and momentum. Differential equations are then formulated from basic physical principles, such as the conservation of energy or momentum. There are a large variety of numerical method that can be used for solving continuum partial differential equation, the most popular being the finite element method (FEM).

The finite element method is a numerical method for approximating a solution to a system of partial differential equations the FEM proceeds by dividing the continuum into a number of elements, each connected to the next by nodes. This discretization process converts the PDE's into a set of coupled ordinary equations that are solved at the nodes of the FE mesh and interpolated throughout the interior of the elements using shape functions. The main advantages of the FEM are its flexibility in geometry, refinement, and loading conditions. It should be noted that the FEM is local, which means that the energy within a body does not change throughout each element and only depends on the energy of the nodes of that element.

The total potential energy under the FE framework consists of two parts, the internal energy U and external work W.

$$\varepsilon_{xx}\varepsilon_{yy}\varepsilon_{zz}\gamma_{xy}\gamma_{yz}\gamma_{xz} \tag{41}$$

The internal energy is the strain energy caused by deformation of the body and can be written as (42):

$$U = \frac{1}{2}\int_{\Omega}\{\sigma\}^{T}\{\varepsilon\}d\Omega = \frac{1}{2}\int_{\Omega}\{\sigma\}^{T}[D]\{\varepsilon\}d\Omega \tag{42}$$

where $\{\sigma\} = \{\sigma_{xx}\sigma_{yy}\sigma_{zz}\tau_{xy}\tau_{yz}\tau_{xz}\}^{T}$ denotes the stress vector, $\{\varepsilon\} = \{\varepsilon_{xx}\varepsilon_{yy}\varepsilon_{zz}\varepsilon_{xy}\varepsilon_{yz}\varepsilon_{xz}\}^{T}$ denotes the strain vector, $[D]$ is the elasticity matrix, and indicate that integration is over the entire domain.

The external work can be written as:

$$W = \int_{\Omega} \{uvw\} \begin{Bmatrix} \mathfrak{I}_x \\ \mathfrak{I}_y \\ \mathfrak{I}_z \end{Bmatrix} d\Omega + \int_{\Gamma} \{uvw\} \begin{Bmatrix} T_x \\ T_y \\ T_z \end{Bmatrix} d\Gamma \tag{43}$$

where u, v and w represent the displacement in x, y, z directions, respectively, is the force vector which contains both applied and body forces, is the surface traction vector, and indicates that the integration of the traction occurs only over the surface of the body. After discretization of the region into a number of elements, point wise discretization of the displacements u, v and z directions, is achieved using shape function $[N]$ for each element, such that the total potential energy becomes:

$$\Pi^g = \frac{1}{2}\{d\}^T \int_{\Omega^g} [B]^T[D][B]\{d\}d\Omega - \{d\}^T \int_{\Omega^g} [N]^T \begin{Bmatrix} \mathfrak{I}_x \\ \mathfrak{I}_y \\ \mathfrak{I}_z \end{Bmatrix} d\Omega - \{d\}^T \int_{\Gamma^g} [N]^T \begin{Bmatrix} T_x \\ T_y \\ T_z \end{Bmatrix} d\Gamma \tag{44}$$

where $[B]$ is the matrix containing the derivation of the shape function, and $[d]$ is a vector containing the displacements.

13.3   CARBON NANO TUBES (CNT): STRUCTURE AND PROPERTIES

A Carbon Nano tube is a tube shaped material, made of carbon, having length-to-diameter ratio over of 100,000,000:1, considerably higher than for any other material. These cylindrical carbon molecules have strange properties such as extraordinary mechanical, electrical properties and thermal conductivity, which are important for different fields of materials science and technology.

Nanotubes are members of the globular shape structural category. They are formed from rolling of one atom thick sheets of carbon, called graphene, which cause to hollow structure. The process of rolling could be done at specific and discrete chiral angles, as nano tube properties are dependent to the rolling angle. The single nanotubes physically line up themselves into ropes situation together by Vander Waals forces.

Chemical bonding in nanotubes describes by orbital hybridization. The chemical bonding of nanotubes is constituted completely ofsp^{2}bonds, similar to those of graphite which are stronger than the sp^{3}bondsfound in diamond, provide nanotubes with their unique strength.

13.3.1 CATEGORIZATION OF NANOTUBES

Carbon nano tube is classified as single walled nano tubes (SWNTs), multi-walled nano tubes (MWNTs) and Double walled carbon nanotubes (DWNTs).

13.3.1.1 SINGLE-WALLED CARBON NANO TUBE

SWNTs have a diameter of near to 1nanometerand tube length of longer than 10^9 times. The structure of a SWNT can be explained by wresting a graphene into a seamless cylinder. The way the graphene is wrested is depicted by a pair of indices *(n, m)*. The integer's *n* and *m* indicate the number of unit vectors in the direction of two points in the honey comb crystal lattice of graphene. If *n=m*, the nanotubes are called *armchair* nanotubes and while *m= 0*, the nanotubes are called *zigzag* nano tubes (Fig. 13.4). If not, they are named *chiral*. The *diameter* of a perfect nano tube can be calculated from its *(n,m)* indices as:

$$d = \frac{a}{\pi}\sqrt{(n^2 + nm + m^2)} = 78.3 \sqrt{((n+m)^2 - nm)} \ \ (pm) \ \ (a= 0.246 \ nm) \quad (45)$$

SWNTs are an important kind of carbon nano tube due to most of their properties change considerably with the *(n, m)* values, and according to *Kataura plot*, this dependence is unsteady. Mechanical properties of single SWNTs were predicted remarkable by Quantum mechanics calculations as Young's modulus of 0.64–1TPa, Tensile Strength of 150–180 GPa, strain to failure of 5–30% while having a relatively low density of 1.4–1.6 g/cm^3.

These high stiffness and superior mechanical properties for SWNTs are due to the chemical structure of the repeat unit. The repeat unit is composed completely of sp^2hybridized carbons and without any points for flexibility or rotation. Also, Single walled nanotubes with diameters of an order of a nanometer can be excellent conductors for electrical industries.

In most cases, SWNTs are synthesized through the reaction of a gaseous carbon feedstock to form the nanotubes on catalyst particles.

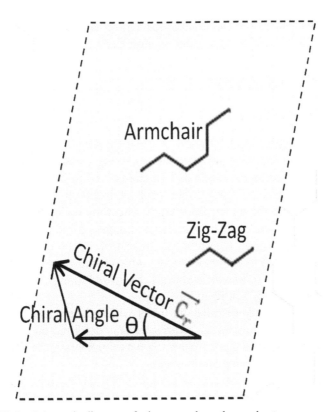

FIGURE 13.4 Schematic diagram of a hexagonal grapheme sheet.

13.3.1.2 MULTI-WALLED CARBON NANO TUBE

Multi-walled nanotubes (MWNT) consist of several concentric graphene tubes, which its interlayer distance is around 3.4Å; close to the distance between graphene layers in graphite. Its single shells in MWNTs can be explained as SWNTs.

13.3.1.3 DOUBLE-WALLED CARBON NANOTUBES (DWNT)

Double-walled carbon nanotubes consist of a particular set of nanotubes because their morphology and properties are comparable to those of SWNT but they have considerably superior chemical resistance. Prominently, it is important when grafting of chemical functions at the surface of the nanotubes is required to achieve new properties of the CNT. In this processing of SWNT, some $C=C$ double bonds will be broken by covalent functionalization, and thus, both mechanical and electrical properties of nanotubes will be modified. About of DWNT, only the outer shell is modified.

13.3.2   CNT PROPERTIES

13.3.2.1   STRENGTH

In terms of strength, tensile strength and elastic modulus are explained and it has not been discovered any material as strong as carbon nanotubes yet. CNTs due to containing the single carbon atoms, having the covalent sp²bonds formed between them and they could resist against high tensile stress. Many studies have been done on tensile strength of carbon nanotubes and totally it was included that single CNT shells have strengths of about 100GPa. Since density of a solid carbon nanotubes is around of 1.3 to 1.4 g/cm³, specific strength of them is up to 48,000 kNm kg⁻¹ which it causes to including carbon nano tube as the best of known materials, compared to high carbon steel that has specific strength of 154 kNm kg⁻¹.

Under extreme tensile strain, the tubes will endure plastic deformation, which means the deformation is invariable. This deformation commences at strains of around 5% and can enhance the maximum strain the tubes undergo before breakage by releasing strain energy.

Despite of the highest strength of single CNT shells, weak shear interactions between near shells and tubes leads to considerable diminutions in the effective strength of multiwalled carbon nanotubes, while crosslink in inner shells and tubes included the strength of these materials is about 60 GPa for multiwalled carbon nanotubes and about 17GPa for double-walled carbon nano tube bundles.

Almost, hollow structure and high aspect ratio of carbon nanotubes lead to their tendency to suffer bending when placed under compressive, torsion stress.

13.3.2.2   HARDNESS

The regular single-walled carbon nanotubes have ability to undergo a transformation to great hard phase and so they can endure a pressure up to 25 GPa without deformation. The bulk modulus of great hard phase nanotubes is around 500 GPa, which is higher than that of diamond (420 GPa for single diamond crystal).

13.3.2.3 ELECTRICAL PROPERTIES

Because of the regularity and exceptional electronic structure of graphene, the structure of a nano tube affects its electrical properties strongly. It has been concluded that for a given *(n, m)* nano tube, while *n=m*, the nano tube is metallic; if *n−m* is a multiple of 3, then the nano tube is semiconducting with a very small band gap, if not, the nano tube is a moderate semiconductor. However, some exceptions are in this rule, because electrical properties can be strongly affected by curvature in small diameter carbon nano tubes. In theory, metallic nanotubes can transmit an electric current density of $4 \times 10 9$ A/cm^2, which is more than 1000 times larger than those of metals such as copper, while electro migration lead to limitation of current densities for copper interconnects.

Because of nanoscale cross-section in carbon nanotubes, electrons spread only along the tube's axis and electron transfer includes quantum effects. As a result, carbon nanotubes are commonly referred to as one-dimensional conductors. The maximum electrical transmission of a SWNT is $2G_0$, where $G_0 = 2e^2/h$ is the transmission of a single ballistic quantum channel.

13.3.2.4 THERMAL PROPERTIES

It is expected that nanotubes act as very good thermal conductors, but they are good insulators laterally to the tube axis. Measurements indicate that SWNTs have a room temperature thermal conductivity along its axis of about 3500 W×m^{-1}K^{-1}; higher than that for copper (385 Wm^{-1}K^{-1}). Also, SWNT has a room temperature thermal conductivity across its axis (in the radial direction) of around 1.52 Wm^{-1}K^{-1}, which is nearly as thermally conductive as soil. The temperature constancy of carbon nanotubes is expected to be up to 2800 °C in vacuum and about 750 °C in air.

13.3.2.5 DEFECTS

As with any material, the essence of a crystallographic defect affects the material properties. Defects can happen in the form of atomic vacancies. High levels of such defects can drop the tensile strength up to 85%. A main example is the Stone Wales defect, which makes a pentagon and heptagon pair by reorganization of the bonds. Having small structure in carbon nanotubes lead to dependency of their tensile strength to the weakest segment.

Also, electrical properties of CNTs can be affected by crystallographic defects. A common result is dropped conductivity through the defective section of the tube. A defect in conductive nanotubes can cause the adjacent section to become semiconducting, and particular monatomic vacancies induce magnetic properties.

Crystallographic defects intensively affect the tube's thermal properties. Such defects cause to phonon scattering, which in turn enhance the relaxation rate of the phonons. This decreases the mean free path and declines the thermal conductivity of nano tube structures. Phonon transport simulations show that alternative defects such as nitrogen or boron will mainly cause to scattering of high frequency optical phonons. However, larger scale defects such as Stone Wales defects lead to phonon scattering over a wide range of frequencies, causing to a greater diminution in thermal conductivity.

13.3.3   METHODS OF CNT PRODUCTION

13.3.3.1   ARC DISCHARGE METHOD

Nanotubes were perceived in 1991 in the carbon soot of graphite electrodes during an arc discharge, by using a current of 100 amps that was intended to create fullerenes. However, for the first time, macroscopic production of carbon nanotubes was done in 1992 by the similar method of 1991. During this process, the carbon included the negative electrode sublimates due to the high discharge temperatures. As the nanotubes were initially discovered using this technique, it has been the most widely used method for synthesis of CNTs. The revenue for this method is up to 30% by weight and it produces both single and multi walled nanotubes with lengths of up to 50 micrometers with few structural defects.

13.3.3.2   LASER ABLATION METHOD

Laser ablation method was developed by Dr. Richard Smalley and co-workers at Rice University. In that time, they were blasting metals with a laser to produce a variety of metal molecules. When they noticed the existence of nanotubes they substituted the metals with graphite to produce multiwalled carbon nanotubes. In the next year, the team applied a composite of graphite and metal catalyst particles to synthesize single walled carbon nano tubes. In laser ablation method, vaporizing a graphite target in a high temperature reactor is done by a pulsed laser while an inert is bled into the chamber. Nanotubes expand on the cooler surfaces of the reactor as the vaporized carbon condenses. A water-cooled surface may be contained in the system to gathering the nanotubes.

The laser ablation method revenues around 70% and produces principally single-walled carbon nanotubes with a controllable diameter determined by the reaction temperature. However, it is more costly than either arc discharge or chemical vapor deposition.

13.3.3.3   PLASMA TORCH

In 2005, a research group from the University of Sherbrooke and the National Research Council of Canada could synthesize Single walled carbon nanotubes by the induction thermal plasma method. This method is alike to arc discharge in that both apply ionized gas to achieve the high temperature necessary to vaporize carbon containing substances and the metal catalysts necessary for the following nano tube development. The thermal plasma is induced by high frequency fluctuating currents in a coil, and is kept in flowing inert gas. Usually, a feedstock of carbon black and metal catalyst particles is supplied into the plasma, and then cooled down to constitute single walled carbon nanotubes. Various single wall carbon nano tube diameter distributions can be synthesized.

The induction thermal plasma method can create up to 2 grams of nano tube material per minute, which is higher than the arc-discharge or the laser ablation methods.

13.3.4   CHEMICAL VAPOR DEPOSITION (CVD)

In 1952 and 1959, the catalytic vapor phase deposition of carbon was studied, and finally, in1993; the carbon nanotubes were constituted by this process. In 2007, researchers at the University of Cincinnati (UC) developed a process to develop aligned carbon nano tube arrays of 18 mm length on a First Nano ET3000 carbon nano tube growth system.

In CVD method, a substrate is prepared with a layer of metal catalyst particles, most usually iron, cobalt, nickel or a combination. The metal nanoparticles can also be formed by other ways, including reduction of oxides or oxides solid solutions. The diameters of the carbon nanotubes, which are to be grown are related to the size of the metal particles. This can be restrained by patterned (or masked) deposition of the metal, annealing, or by plasma etching of a metal layer. The substrate is heated to around of 700 °C. To begin the enlargement of nanotubes, two types of gas are bled into the reactor: a process gas (such as ammonia, nitrogen or hydrogen) and a gas containing carbon (such as acetylene, ethylene, ethanol or methane). Nanotubes develop at the sites of the metal catalyst; the gas containing carbon is broken apart at the surface of the catalyst particle, and the carbon is transferred to the edges of the particle, where it forms the nano tubes. The catalyst particles can stay at the tips of the growing nano tube during expansion, or remain at the nano tube base, depending on the adhesion between the catalyst particle and the substrate.

CVD is a general method for the commercial production of carbon nanotubes. For this idea, the metal nanoparticles are mixed with a catalyst support such as MgO or Al_2O_3 to enhance the surface area for higher revenue of the catalytic reaction of the carbon feedstock with the metal particles. One matter in this synthesis method is the removal of the catalyst support via an acid treatment, which sometimes could destroy the primary structure of the carbon nanotubes. However, other catalyst supports that are soluble in water have verified effective for nano tube development.

Of the different means for nano tube synthesis, CVD indicates the most promise for industrial scale deposition, due to its price/unit ratio, and because CVD is capable of increasing nanotubes directly on a desired substrate, whereas the nanotubes must be collected in the other expansion techniques. The development sites are manageable by careful deposition of the catalyst.

13.3.5   SUPER-GROWTH CVD

Kenji Hata, Sumio Iijima and co-workers at AIST, Japan; developed super growth CVD (water assisted chemical vapor deposition), by adding water into CVD reactor to improve the activity and lifetime of the catalyst. Dense millimeter all nano tube *forests*, aligned normal to the substrate, were created. The *forests* expansion rate could be extracted, as:

$$H(t) = \beta\tau_0(1 - e^{-\frac{t}{\tau_0}})$$

(46)

In this equation, β is the initial expansion rate andis the characteristic catalyst lifetime.

13.4   SIMULATION OF CNT'S MECHANICAL PROPERTIES

13.4.1   MODELING TECHNIQUES

The theoretical efforts in simulation of CNT mechanical properties can be categorized in three groups as follow:
- Atomistic simulation

- Continuum simulation
- Nano-scale continuum modeling

13.4.1.1 ATOMISTIC SIMULATION

Based on interactive forces and boundary conditions, atomistic modeling predicts the positions of atoms. Atomistic modeling techniques can be classified into three main categories, namely the MD, MC and AB initio approaches. Other atomistic modeling techniques such as tight bonding molecular dynamics (TBMD), local density (LD), density functional theory (DFT), Morse potential function model, and modified Morse potential function model were also applied as discussed in last session.

The first technique used for simulating the behaviors of CNTs was MD method. This method uses realistic force fields (many-body interatomic potential functions) to determination the total energy of a system of particles. Whit the calculation of the total potential energy and force fields of a system, the realistic calculations of the behavior and the properties of a system of atoms and molecules can be acquired. Although the main aspect of both MD and MC simulations methods is based on second Newton's law, MD methods are deterministic approaches, in comparison to the MC methods that are stochastic ones.

In spite the MD and MC methods depend on the potentials that the forces acting on atoms by differentiating inter atomic potential functions, the AB initio techniques are accurate methods which are based on an accurate solution of the Schrödinger equation. Furthermore the AB initio techniques are potential-free methods wherein the atoms forces are determined by electronic structure calculations progressively.

In generally, MD simulations provide good predictions of the mechanical properties of CNTs under external forces. However, MD simulations take long times to produce the results and consumes a large amount of computational resources, especially when dealing with long and multiwalled CNTs in corpora ting a large number of atoms.

13.4.1.2 CONTINUUM MODELING

Continuum mechanics-based models are used by many researches to investigate properties of CNTs. The basic assumption in these theories is that a CNT can be modeled as a continuum structure which has continuous distributions of mass, density, stiffness, etc. So, the lattice structure of CNT is simply neglected in and it is replaced with a continuum medium. It is important to meticulously investigate the validity of continuum mechanics approaches for modeling CNTs, which the real discrete nano-structure of CNT is replaced with a continuum one. The continuum modeling can be either accomplished analytical (micromechanics) or numerically representing FEM.

Continuum shell models used to study the CNT properties and showed similarities between MD simulations of macroscopic shell model. Because of the neglecting the discrete nature of the CNT geometry in this method, it has shown that mechanical properties of CNTs were strongly dependent on atomic structure of the tubes and like the curvature and chirality effects, the mechanical behavior of CNTs cannot be calculated in an isotropic shell model. Different from common *shell model,* which is constructed as an isotropic continuum shell with constant elastic properties for SWCNTs, the MBASM model can predict the chirality induced anisotropic effects on some mechanical behaviors of CNTs by incorporating molecular and continuum mechanics solutions. One of the other theory is shallow shell theories, this theory are not accurate for CNT analysis because of CNT is a nonshallow structure. Only more complex shell is capable of reproducing the results of MD simulations.

Some parameters, such as wall thickness of CNTs are not well defined in the continuum mechanics. For instance, value of 0.34 nm which is inter planar spacing between graphene sheets in graphite is widely used for tube thickness in many continuum models.

The finite element method works as the numerical methods for determining the energy minimizing displacement fields, while atomistic analysis is used to determine the energy of a given configuration. This is in contrast to normal finite element approaches, where the constitutive input is made via phenomenological models. The method is successful in capturing the structure and energetic of dislocations. Finite element modeling is directed by using 3D beam element, which is as equivalent beam to construct the CNT. The obtained results will be useful in realizing interactions between the nanostructures and substrates and also designing composites systems.

13.4.1.3   NANO-SCALE CONTINUUM MODELING

Unlike to continuum modeling of CNTs where the entirely discrete structure of CNT is replaced with a continuum medium, nano-scale continuum modeling provides a rationally acceptable compromise in the modeling process by replacing C–C bond with a continuum element. In the other hand, in nano-scale continuum modeling the molecular interactions between C–C bonds are captured using structural members whose properties are obtained by atomistic modeling. Development of nano-scale continuum theories has stimulated more excitement by incorporating continuum mechanics theories at the scale of nano. Nano-scale continuum modeling is usually accomplished numerically in the form of finite element modeling. Different elements consisting of rod, truss, spring and beam are used to simulate C–C bonds. The two common method of nano-scale continuum are quasi-continuum and equivalent-continuum methods, which have been used in nano-scale continuum modeling.

The *quasi-continuum (QC)* method which presents a relationship between the deformations of a continuum with that of its crystal lattice uses the classical Cauchy-Born rule and representative atoms. The quasi-continuum method mixes atomistic-continuum formulation and is based on a finite element discretization of a continuum mechanics variation principle.

The *equivalent-continuum method* developed by providing a correlation between computational chemistry and continuum structural mechanics. It has considered being equal total molecular potential energy of a nanostructure with the strain energy of its equivalent continuum elements.

This method has been proposed for developing structure-property relationships of nano-structured materials and works as a link between computational chemistry and solid mechanics by substituting discrete molecular structures with equivalent-continuum models. It has been shown that this substitution may be accomplished by equating the molecular potential energy of a nano-structured material with the strain energy of representative truss and continuum models.

Because of the approach uses the energy terms that are associated with molecular mechanics modeling, a brief description of molecular mechanics is given first followed by an outline of the equivalent-truss and equivalent-continuum model development.

13.4.1.3.1 MOLECULAR MECHANICS

An important part in molecular mechanics calculations of the nano-structure materials is the determination of the forces between individual atoms. This description is characterized by a force field. In the most general form, the total molecular potential energy, E, for a nano-structured material is described by the sum of many individual energy contributions:

$$E_{total} = \sum E_\rho + \sum E_\theta + \sum E_\tau + \sum E_\omega + \sum E_{vdw} + \sum E_{el} \qquad (47)$$

where , , , are the energies associated with bond stretching, angle variation, torsion, and inversion, respectively. The atomic deformation mechanisms are illustrated in Figs. 13.5 and 13.6.

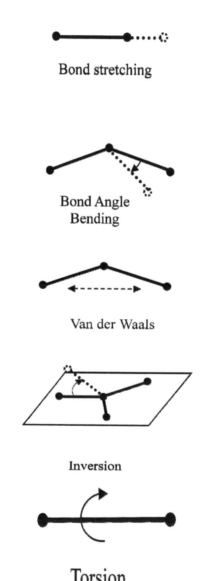

Bond stretching

**Bond Angle
Bending**

Van der Waals

Inversion

Torsion

FIGURE 13.5   Atomistic bond interaction mechanisms.

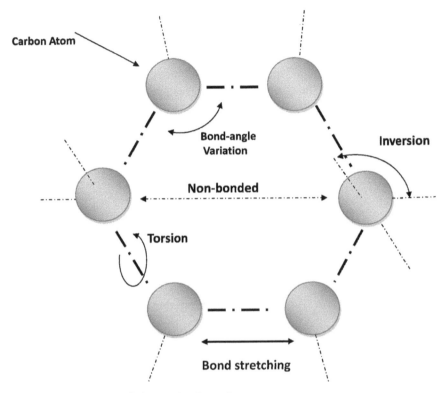

FIGURE 13.6   Schematic interaction for carbon atoms.

The non-bonded interaction energies consist of Vander Waals and electrostatic, terms. Depending on the type of material and loading conditions various functional forms may be used for these energy terms. In condition where experimental data are either unavailable or very difficult to measure, quantum mechanical calculations can be a source of information for defining the force field.

In order to simplify the calculation of the total molecular potential energy of complex molecular structures and loading conditions, the molecular model substituted by intermediate model with a pin-jointed Truss model based on the nature of molecular force fields, to represent the energies given by Eq. (41), where each truss member represents the forces between two atoms.

So, a truss model allows the mechanical behavior of the nano-structured system to be accurately modeled in terms of displacements of the atoms.

This mechanical representation of the lattice behavior serves as an intermediate step in linking the molecular potential with an equivalent-continuum model.

In the truss model, each truss element represented a chemical bond or a nonbonded interaction. The stretching potential of each bond corresponds with the stretching of the corresponding truss element. Atoms in a lattice have been viewed as masses that are held in place with atomic forces that is similar to elastic springs. Therefore, bending of Truss elements is not needed to simulate the chemical bonds, and it is assumed that each truss joint is pinned, not fixed, Fig. 13.7 shown the atomistic-based continuum modeling and RVE of the chemical, truss and continuum models.

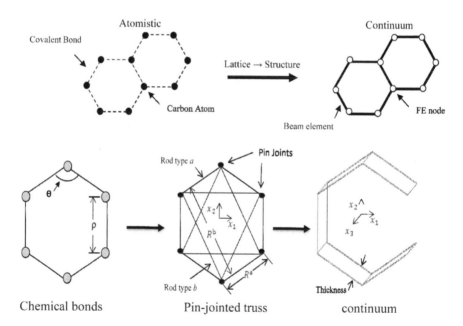

FIGURE 13.7   Atomistic-based continuum modeling and RVE of the chemical, truss and continuum models.

The mechanical strain energy, , of the truss model is expressed in the form:

$$\Lambda^t = \sum_n \sum_m \frac{A_m^n Y_m^n}{2 R_m^n} (r_m^n - R_m^n)^2 \tag{48}$$

where and are the cross-sectional area and Young's modulus, respectively, of rod m of truss member type n. The term is the stretching of rod m of truss member type n, where and are the undeformed and deformed lengths of the truss elements, respectively.

In order to represent the chemical behavior with the truss model, Eq. (47) must be consider being equal with Eq. (48) in a physically meaningful manner. Each of the two equations is sums of energies for particular degrees of freedom. In comparison of the Eq. (47) that have bond angle variance and torsion terms the Eq. (48) has stretching term only, so it made the main difficulty for substitute these tow equations. No generalization can be made for overcoming this difficulty for every nano-structured system. It means that possible solution must be determined for a specific nano-Structured material depending on the geometry, loading conditions, and degree of accuracy sought in the model.

13.5   LITERATURE REVIEW ON CNT SIMULATION

Different researchers have been doing many efforts to simulate mechanical properties of CNT, in generally the main trends of these methods employed by different researchers to predict the elastic modulus of SWCNTs results in terms of three main parameters of morphology: radius, chirality, and wall thickness. The dependency of results to the diameter of CNT becomes less pronounced when nonlinear inter atomic potentials are employed instead of linear ones.

There are plenty of experimental and theoretical techniques for characterizing Young's modulus of carbon nanotubes, all three main group of CNT modeling; MD, continuum modeling (CM) and nano-scale continuum modeling (NCM) have been used in the literature. Some of recent different theoretical methods for predicting Young's modulus of carbon nanotubes are summarized in (Tables 13.1–13.3).

TABLE 13.1   MD Methods for Prediction Young' Modulus of CNT

Researchers	Year	Method	Young's Modulus (TPa)	Results
Liew et al. [1]	2004	Second generation reactive of empirical bond-order (REBO)	1.043	Examining the elastic and plastic properties of CNTs under axial tension
H. W. Zhang [2]	2005	Modified Morse Potentials and Tersoff–Brenner potential	1.08	Predicting the elastic properties of SWCNTs based on the classical Cauchy–Born rule
Agrawal et al. [3]	2006	A combination of a second generation reactive empirical bond order potential and vdW interactions	0.55, 0.73, 0.74, 0.76	Predicting Young's modulus by four MD approaches for an armchair (14,14) type SWCNT and investigating effect of defects in the form of vacancies, van der Waals (vdW) interactions, chirality, and diameter
Cheng et al. [4]	2008	MD simulations using Tersoff–Brenner potential to simulate covalent bonds while using Lennard–Jones to model interlayer interactions	1.4 for armchair 1.2 for zigzag	Evaluating the influence of surface effect resulting in relaxed unstrained deformation and in-layer nonbonded interactions using atomistic continuum modeling approach
Cai et al. [5]	2009	Linear scaling self-consistent charge, density functional tight binding (SCC-DFTB) and an ab initio Dmol³	—	Investigating the energy and Young's modulus as a function of tube length for (10, 0) SWCNTs
Ranjbartoreh and Wang[6]	2010	Large-scale atomic/molecular massively parallel simulator (LAMMPS) code	0.788 for armchair 1.176 for zigzag	Effects of chirality and Van der Waals interaction on Young's modulus, elastic compressive modulus, bending, tensile, compressive stiffness, and critical axial force of DWCNTs

TABLE 13.2   Continuum Simulation Methods for Prediction Young' Modulus of CNT

Researchers	Year	Method	Young's Modulus (TPa)	Results
Sears and Batra [7]	2004	Equivalent continuum tube	2.52	Results of the molecular-mechanics simulations of a SWNT is used to derive the thickness and values of the two elastic modulus of an isotropic linear elastic cylindrical tube equivalent to the SWNT
Wang [8]	2004	Stretching and rotating springs, Equivalent continuum plate	0.2–0.8 for zigzags 0.57–0.54 for armchairs	Effective in-plane stiffness and bending rigidity of SWCNTs
Kalamkarov et al. [9]	2006	Asymptotic homogenization, cylindrical network shell	1.71	Young's modulus of SW-CNTs
Gupta and Batra [10]	2008	Equivalent continuum structure	0.964 ± 0.035	Predicting Young's modulus of SWCNTs b equating frequencies of axial and torsional modes of vibration of the ECS to those of SWCNT computed through numerical simulations using the MM3 potential
Giannopoulos et al. [11]	2008	FEM, spring elements	1.2478	Young's and Shear modulus of SWCNTs using three dimensional spring-like elements
Papanikos et al. [12]	2008	3-D beam element	0.4–2.08	3D FE analysis, assuming a linear behavior of the C–C bonds

TABLE 13.3   Nano-scale Continuum Methods for Prediction Young' Modulus of CNT

Researchers	Year	Method	Young's Modulus (TPa)	Results
Li and Chou[13]	2003	Computational model using beam element and non-linear truss rod element	1.05 ± 0.05	Studying of elastic behavior of MWCNTs and investigating the influence of diameter, chirality and the number of tube layers on elastic modulus
Natsuki et al. [14]	2004	Frame structure model with a spring constant for axial deformation and bending of C–C bond using analytical method	0.61–0.48	A two dimensional continuum-shell model which is composed of the discrete molecular structures linked by the carbon–carbon bonds
Natsuki and Endo [15]	2004	Analytical analysis by exerting nonlinear Morse potential	0.94 for armchair (10,10)	Mechanical properties of armchair and zigzag CNT are investigated. The results show the atomic structure of CNT has a remarkable effect on stress– strain behavior
Xiao et al. [16]	2005	Analytical molecular structural mechanics based on the Morse potential	1–1.2	Young's modulus of SWCNTs under tension and torsion and results are sensitive to the tube diameter and the helicity
Tserpes and Papanikos [17]	2005	FE method employing 3D elastic beam element	for chirality(8,8) and thickness 0.147 nm 2.377	Evaluating Young's modulus of SWCNT using FEM and investigating the influence of wall thickness, tube diameter and chirality on CNT Young's modulus
Jalalahmadi and Naghdabadi [18]	2007	Finite element modeling employing beam element based on Morse potential	3.296, 3.312, 3.514	Predicting Young's modulus utilizing FEM and Morse potential to obtain mechanical properties of beam elements, Moreover investigating of wall thickness, diameter and chirality effects on Young's modulus of SWCNT
Meo and Rossi [19]	2007	Nonlinear and torsion springs	0.912 for zigzag structures 0.920 for armchair structures	Predicting the ultimate strength and strain of SWCNTs and effects of chirality and defections

TABLE 13.3   *(Continued)*

Researchers	*Year*	*Method*	*Young's Modulus (TPa)*	*Results*
PourAkbar Saffar et al. [20]	2008	Finite element method utilizing 3D beam element	1.01 for (10,10)	Obtaining Young modulus of CNT to use it in FE analysis and investigating Young modulus of CNT reinforced composite
Cheng et al. [21]	2008	MD simulations using Tersoff–Brenner potential to simulate covalent bonds while using Lennard–Jones to model interlayer interactions. Finite element analysis employing nonlinear spring element and equivalent beam element to model in-layer non-bonded interactions and covalent bond of two neighbor atoms, respectively	1.4 for armchair and 1.2 for zigzag	Evaluating the influence of surface effect resulting in relaxed unstrained deformation and in-layer nonbonded interactions using atomistic continuum modeling approach
Avila and Lacerda [22]	2008	FEM using 3D beam element to simulate C–Cbond	0.95–5.5 by altering the CNT radius Constructing SWCNT (armchair, zigzag and chiral)	model based on molecular mechanic and evaluating its Young's modulus and Poisson's ratio
Wernik and Meguid [23]	2010	FEM using beam element to model the stretching component of potential and also nonlinear rotational spring to take account the angle-bending component	0.9448 for armchair (9, 9)	Investigating nonlinear response of CNT using modified Morse potential, also studying the fracture process under tensile loading and torsional bucking

TABLE 13.3   *(Continued)*

Researchers	Year	Method	Young's Modulus (TPa)	Results
Shokrieh and Rafiee [24]	2010	Linking between inter-atomic potential energies of lattice molecular structure and strain energies of equivalent discrete frame structure	1.033–1.042	A closed-form solution for prediction of Young's modulus of a graphene sheet and CNTs and finite element modeling of CNT using beam element
Lu and Hu [25]	2012	Using nonlinear potential to simulate C–C bond with considering elliptical-like cross section area of C–C bond	For diameter range 0.375–1.8, obtained Young's modulus is 0.989–1.058	Predicting mechanical properties of CNT using FEM, also investigating rolling energy per atom to roll graphene sheet to SWNT
Rafiee and Heidarhaei [26]	2012	Nonlinear FEM using nonlinear springs for both bond stretching and bond angle variations	1.325	Young's modulus of SW-CNTs and investigating effects of chirality and diameter on that

13.5.1   BUCKLING BEHAVIOR OF CNTS

One of the technical applications on CNT related to buckling properties is the ability of nanotubes to recover from elastic buckling, which allows them to be used several times without damage. One of the most effective parameters on buckling behaviors of CNT on compression and torsion is chirality that thoroughly investigated by many researchers [27, 28]. Chang et al. [29] showed that zigzag chirality is more stable than armchair one with the same diameter under axial compression. Wang et al. [30] reviewed buckling behavior of CNTs recently based on special characteristics of buckling behavior of CNTs.

The MD simulations have been used widely for modeling buckling behavior of CNTs [31–35]. MD use simulation package, which included Newtonian equations of motion based on Tersoff-Brenner potential of inter atomic forces. Yakobson et al. [36] used MD simulations for buckling of SWCNTs and showed that CNTs can provide extreme strain without permanent deformation or atomic rearrangement. Molecular dynamics simulation by destabilizing load that composed of axial compression, torsion have been used to buckling and postbucking analysis of SWCNT [37].

The buckling properties and corresponding mode shapes were studied under different rotational and axial displacement rates which indicated the strongly dependency of critical loads and buckling deformations to these displacement rates. Also buckling responses of multiwalled carbon nanotubes associated to torsion springs in electromechanical devices were obtained by Jeong [38] using classical MD simulations.

Shell theories and molecular structural simulations have been used by some researchers to study buckling behavior and CNT change ability for compression and torsion. Silvestre et al. [39] showed the inability of Donnell shell theory [40] and shown that the Sanders shell theory [41] is accurate in reproducing buckling strains and mode shapes of axially compressed CNTs with small aspect ratios. It's pointed out that the main reason for the incorrectness of Donnell shell model is the inadequate kinematic hypotheses underlying it.

It is exhibited that using Donnell shell and uniform helix deflected shape of CNT simultaneously, leads to incorrect value of the critical angle of twist, conversely Sanders shell model with non uniform helix deflected shape presents correct results of critical twist angle. Besides, Silvestre et al. [42] presented an investigation on linear buckling and postbuckling behavior of CNTs using molecular dynamic simulation under pure shortening and twisting.

Ghavamian and Öchsner [43] were analyzed the effect of defects on the buckling behavior of single and multi-walled CNT based on FE method considering three most likely atomic defects including impurities, vacancies (carbon vacancy) and introduced disturbance. The results demonstrate that the existence of any type kinks in CNTs structure, conducts to lower critical load and lower buckling properties.

Zhang et al. [44] performed an effort on the accuracy of the Euler Bernoulli beam model and Donnell shell model and their nonlocal counterparts in predicting the buckling strains of single-walled CNTs. Comparing with MD simulation results, they concluded that the simple Euler-Bernoulli beam model is sufficient for predicting the buckling strains of CNTs with large aspect ratios (i.e., length to diameter ratio $L/d > 10$). The refined Timoshenko's beam model for nonlocal beam theory is needed for CNTs with intermediate aspect ratios (i.e., $8 < L/d < 10$). The Donnell thin shell theory is unable to capture the length dependent critical strains obtained by MD simulations for CNTs with small aspect ratios (i.e., $L/d < 8$) and hence this simple shell theory is unable to model small aspect ratio CNTs. (Tables 13.4–13.6) summarizes some of theoretical simulations of buckling behavior of CNTs.

TABLE 13.4 MD Simulation of CNT Buckling

Researchers	Year	Method	Chairality	Length (nm)	Diameter (nm)	Results
Wang et al. [30]	2005	Tersoff–Brenner potential	(n,n)(n,0)	7–19	0.5–1.7	Obtaining critical stresses and comparing between the buckling behavior in nano and macroscopic scale
Sears and Batra [45]	2006	MM3 class II pair wise potential, FEM, equivalent continuum structures using Euler buckling theory	Various zig-zag CNTs	50-350Å	---	Buckling of axially compressed multi-walled carbon nanotubes by using molecular mechanics simulations and developing continuum structures equivalent to the nanotubes

Xin et al. [46]	2007	Molecular dynamics simulation using Morse potential, harmonic angle potential and Lennard–Jones potential	(7,7) and (12,0)	---	---	Studying buckling behavior of SWCNTs under axial compression based on MD method, also evaluating the impression of tube length, temperature and configuration of initial defects on mechanical properties of SWCNTs
Silvestre et al. [39]	2011	The second generation reactive empirical bond order (REBO) potential	(5, 5)(7,7)	---	---	Comparison between buckling behavior of SWCNTs with Donnell and Sanders shell theories and MD results
Ansari et al. [35]	2011	"AIREBO" potential	(8,8)(14, 0)	---	---	Axial buckling response of SWCNTs based on nonlocal elasticity continuum with different BCs and extracting nonlocal
Silvestre et al. [47]	2012	Tresoff-Bernner Covalent Potential	(8,8) (5,5) (6,3)	---	---	Buckling behavior of the CNTs under Pure shortening and Pure twisting as well as their pre-critical and post-critical stiffness

TABLE 13.5   Continuum Simulation of CNT Buckling

Researchers	Year	Method	Chairality	Length (nm)	Diameter (nm)	Results
He et al. [48]	2005	Continuum cylindrical shell	---			Establishing an algorithm for buckling analysis of multi-walled CNTs based on derived formula which considering Van der Waals interaction between any two layers of MWCNT
Ghorbanpour Arani et al.[49]	2007	FEM, cylindrical shell	---	11.2 and 44.8	3.2	Pure axially compressed buckling and combined loading effects on SW-CNT
Yao et al. [50]	2008	FEM, elastic shell theory	---	---	0.5–3	Bending buckling of single- double- and multi walled CNTs
Guo et al. [51]	2008	Atomic scale finite element method	(15,0) (10,0)	8.3 and 8.38	---	Bending buckling of SWCNTs
Chan et al [52]	2011	Utilizing Donnell shell equilibrium equation and also Euler–Bernoulli beam equation incorporating curvature effect	Various chairality	---	---	Investigating pre and post-buckling behavior of MW-CNTs and multi-walled carbon nano peapods considering vdW interactions between the adjacent walls of the CNTs and the interactions between the fullerenes and the inner wall of the nanotube

TABLE 13.5 *(Continued)*

| Silvestre [53] | 2012 | Donnell and Sanders shell model with non-uniform helix deflected shape | Various length to diameter ratio | Various length to diameter ratio | Investigating of critical twist angle of SWCNT and comparing the accuracy of twoDonnell and Sanders shell model withuniform and non-uniform helix deflectedshape |

TABLE 13.6 Nano-scale Continuum Simulation of CNT Buckling

Researchers	*Year*	*Method*	*Chairality*	*Length (nm)*	*Diameter (nm)*	*Results*
Li and Chou[54]	2004	Molecular structural mechanics with 3D space frame-like structures with beams	(3,3)(8,8) (5,0)(14,0)	---	0.4–1.2	Buckling behaviors under either compression or bending for both single-and double-walled CNTs
Chang et al.[29]	2005	Analytical molecular mechanics, potential energy	---	---	1–5	Critical buckling strain under axial compression
Hu et al. [55]	2007	Molecular structural mechanics with 3D beam elements	Various type of armchair and zigzag	Various length to diameter ratio	Various length to diameter ratio	Investigating of buckling characteristics of SW-CNT and DW-CNT by using the beam element to model C–C bond and proposed rod element to model vdW forces in MWCNT, also the validity of Euler's beam buckling theory and shell buckling mode are studied

13.5.2   VIBRATIONS ANALYSIS

Dynamic mechanical behaviors of CNTs are of importance in various applications, such as high frequency oscillators and sensors [56]. By adding CNTs to polymer the fundamental frequencies of CNT reinforced polymer can be improved remarkably without significant change in the mass density of material [57]. It is importance indicate that the dynamic mechanical analysis confirms strong influence of CNTs on the composite damping properties [58].

The simulation methods of CNT's vibrating were reviewed by Gibson et al. [59] in 2007. Considering wide applications of CNTs, receiving natural frequencies and mode shapes by assembling accurate theoretical model. For instance, the oscillation frequency is a key property of the resonator when CNTs are used as nano mechanical resonators. Moreover, by using accurate theoretical model to acquire natural frequencies and mode shapes, the elastic modulus of CNTs can be calculated indirectly.

Timoshenko's beam model used by Wang et al. [60] for free vibrations of MWCNTs study; it was shown that the frequencies are significantly over predicted by the Euler's beam theory when the aspect ratios are small and when considering high vibration modes. They indicated that the Timoshenko's beam model should be used for a better prediction of the frequencies especially when small aspect ratio and high vibration modes are considered.

Hu et al. presented a review of recent studies on continuum models and MD simulations of CNT's vibrations briefly [61]. Three constructed model of SWCNT consisting of Timoshenko's beam, Euler-Bernoulli's beam and MD simulations are investigated and results show that fundamental frequency decreases as the length of a SWCNT increases and also the Timoshenko's beam model provides a better prediction of short CNT's frequencies than that of Euler-Bernoulli beam's model. Comparing the fundamental frequency results of transverse vibrations of cantilevered SWCNTs it can be seen that both beam models are not able to predict the fundamental frequency of cantilevered SWCNTs shorter than 3.5 nm.

An atomistic modeling technique and molecular structural mechanics are used by Li and Chou [62] to calculated fundamental natural frequency of SWCNTs. A free and forced vibrations of SWCNT have been assessed by Arghavan and Singh [63]using space frame elements with extensional, bending and torsion stiffness properties to modeling SWCNTs and compared their results with reported results by of Sakhaee pour [64] and Li and Chou [62]. Their results were in close agreement with two other results, within three to five percent.

Furthermore, free vibrations of SWCNTs have been studied by Gupta et al. [65] using the MM3 potential. They mentioned that calculations based on modeling SWCNTs as a beam will overestimate fundamental frequencies of the SWCNT. They derived the thickness of a SWCNT/shell to compare the frequencies of a SWCNT obtained by MM simulations with that of a shell model and provided an expression for the wall thickness in terms of the tube radius and the bond length in the initial relaxed configuration of a SWCNT. Aydogdu [66] showed that axial vibration frequencies of SWCNT embedded in an elastic medium highly over estimated by the classical rod model because of ignoring the effect of small length scale and developed an elastic rod model based on local and nonlocal rod theories to investigate the small scale effect on the axial vibrations of SWCNTs.

The vibration properties of two and three junctioned of carbon nanotubes considering different boundary conditions and geometries were studied by Fakhrabadi et al. [67]. The results show that the tighter boundaries, larger diameters and shorter lengths lead to higher natural frequencies. Lee and Lee [68] fulfilled modal analysis of SWCNT's and nano cones (SWCNCs) using finite element method with ANSYS commercial package. The vibration behaviors of SWCNT with fixed beam and cantilever boundary conditions with different cross-section types consisting of circle and ellipse were constructed using 3D elastic beams and point masses. The nonlinear vibration of an embedded CNT's were studied by Fu et al. [69] and the results reveal that the nonlinear free vibration of nano-tubes is effected significantly by surrounding elastic medium.

The same investigation has been accomplished by Ansari and Hem-matnezhad [70] using the variational iteration method (VIM). Wang et al. [71] have studied axisymmetric vibrations of SWCNT immerged in water which in contrast to solid liquid system; a submerged SWCNT is coupled with surrounding water via vdW interaction. The analysis of DWCNTs vibration characteristics considering simply support boundary condition are carried out by Natsuki et al. [72] based on Euler-Bernoulli's beam theory. Subsequently, it was found that the vibration modes of DWCNTs are non coaxial inter tube vibrations and deflection of inner and outer nano tube can come about in same or opposite deflections.

More-over the vibration analysis of MWCNTs were implemented by Aydogdu [73] using generalized shear deformation beam theory (GSDBT). Parabolic shear deformation theory (PSDT) was used in the specific solutions and the results showed remarkable difference between PSDT and Euler beam theory and also the importance of vdW force presence for small inner radius. Lei et al. [74] have presented a theoretical vibration analysis of the radial breathing mode (RBM) of DWCNTs subjected to pressure based on elastic continuum model. It was shown that the frequency of RBM increases perspicuously as the pressure increases under different conditions.

The influences of shear deformation, boundary conditions and vdW coefficient on the transverse vibration of MWCNT were studied by Ambrosini and Borbón [75, 76]. The results reveal that the noncoaxial inter tube frequencies are independent of the shear deformation and the boundary conditions and also are strongly influenced by the vdW coefficient used. The performed investigations on the simulation of vibrations properties classified by the method of modeling are presented in Tables 13.7–13.9. As it can be seen 5–7, 5–8, and 5–9, the majority of investigations used continuum modeling and simply replaced a CNT with hollow thin cylinder to study the vibrations of CNT. This modeling strategy cannot simulate the real behavior of CNT, since the lattice structure is neglected. In other word, these investigations simply studied the vibration behavior of a continuum cylinder with equivalent mechanical properties of CNT. Actually nano-scale continuum modeling is preferred for investigating vibrations of CNTs, since it was reported that natural frequencies of CNTs depend on both chirality and boundary conditions.

TABLE 13.7   MD Methods for Prediction Vibrational Properties of CNT

Researchers	*Year*	*Method*	*Chairality*	*Aspect Ratio (L/D)*	*Results*
Gupta and Batra[77]	2008	MM3 potential	Nineteen armchair, zigzag, and chiral SWCNTs have been discussed	15	Axial, torsion and radial breathing mode (RBM) vibrations of free–free unstressed SWCNTs and identifying equivalent continuum structure
Gupta et al.[65]	2010	MM3 potential	Thirty-three armchair, zig-zag band chiral SWCNTs have been discussed	3–15	Free vibrations of free end SWCNTs and identifying equivalent continuum structure of those SWCNTs
Ansari et al.[78]	2012	Adaptive Intermolecular Reactive Empirical Bond Order (AIREBO) potential	(8,8)	8.3–39.1	Vibration characteristics and comparison between different gradient theories as well as different beam assumptions in predicting the free vibrations of SWCNTs

TABLE 13.7 *(Continued)*

Researchers	Year	Method	Chairality	Aspect Ratio (L/D)	Results
Ansari et al.[79]	2012	Tersoff–Brenner and Lennard–Jones potential	(5,5)(10,10) (9,0)	3.61–14.46	Vibrations of single- and double-walled carbon nanotubes under various layer-wise boundary condition

TABLE 13.8 Continuum Methods for Prediction Vibration Properties of CNTs

Researchers	Year	Method	Chairality	Aspect Ratio (L/D)	Results
Wang et al.[60]	2006	Timoshenko beam theory	0	10, 30, 50, 10	Free vibrations of MWCNTs using Timoshenko beam
Sun and Liu[80]	2007	Donnell's equilibrium equation	---	---	Vibration characteristics of MWCNTs with initial axial loading
Ke et al.[81]	2009	Eringen's nonlocal elasticity theory and von Karman geometric nonlinearity using nonlocal Timoshenko beam	---	10, 20, 30, 40	Nonlinear free vibration of embedded double-walled CNTs at different boundary condition
Ghavanloo and Fazelzadeh[82]	2012	Anisotropic elastic shell using Flugge shell theory	Twenty-four armchair, zigzag and chiral SWCNTs have been discussed	---	Investigating of free and forced vibration of SWCNTs including chirality effect
Aydogdu Metin[66]	2012	Nonlocal elasticity theory, an elastic rod model	---	---	Axial vibration of single walled carbon nanotube embedded in an elastic medium and investigating effect of various parameters like stiffness of elastic medium, boundary conditions and nonlocal parameters on the axial vibration

TABLE 13.8   *(Continued)*

Researchers	Year	Method	Chairality	Aspect Ratio (L/D)	Results
Khosrozadeh and Haja-basi[83]	2012	Nonlocal Euler–Bernoulli beam theory	---	---	Studying of nonlinear free vibration of DWCANTs considering nonlinear inter-layer Van der Waals interactions, also discussing about nonlocal

TABLE 13.9   Nano-scale Continuum Modeling Simulation for Prediction Vibration Properties of CNTs

Researchers	Year	Method	Chairality	Aspect Ratio (L/D)	Results
Georgantzinos et al.[84]	2008	Numerical analysis based on atomistic micro-structure of nano-tube by using linear spring element	Armchair and zigzag	Various aspect ratio	Investigating vibration analysis of single-walled carbon nanotubes, considering different support conditions and defects
Georgantzinos and Anifantis[85]	2009	FEM, based on linear nano-springs to simulate the interatomic behavior	Armchair and zigzag chiralities	3–20	Obtaining mode shapes and natural frequencies of MWCNTs as well as investigating the influence of van der Waals interactions on vibration charac-teristics at differ-ent BCs
Sakhaee Pour et al.[64]	2009	FEM using 3D beam element and point mass	Zigzag and armchair	Various aspect ratio	Computing natural frequencies of SWCNT using atomistic simula-tion approach with considering bridge and cantilever like boundary condi-tions

13.5.3 SUMMARY AND CONCLUSION ON CNT SIMULATION

The CNTs modeling techniques can be classified into three main categories of atomistic modeling, continuum modeling and nano-scale continuum modeling. The atomistic modeling consists of MD, MC and AB initio methods. Both MD and MC methods are constructed on the basis of second Newton's law. While MD method deals with deterministic equations, MC is a stochastic approach. Although both MD and MC depend on potential, AB initio is an accurate and potential free method relying on solving Schrödinger equation. Atomistic modeling techniques are suffering from some shortcomings which can be summarized as: (i) inapplicability of modeling large number of atoms (ii) huge amount of computational tasks (iii) complex formulations. Other atomistic methods such as tight bonding molecular dynamic, local density, density functional theory, Morse potential model and modified Morse potential model are also available which are in need of intensive calculations.

On the other hands, continuum modeling originated from continuum mechanics are also applied to study mechanical behavior of CNTs. Comprising of analytical and numerical approaches, the validity of continuum modeling has to be carefully observed where in lattice structure of CNT is replaced with a continuum medium. Numerical continuum modeling is accomplished through finite element modeling using shell or curved plate elements. The degree to which this strategy, that is, neglecting lattice structure of CNT, will lead us to sufficiently accurate results is under question. Moreover, it was extensively observed that almost all properties of CNTs (mechanical, buckling, vibrations and thermal properties) depend on the chirality of CNT; thus continuum modeling cannot address this important issue.

Recently, nano-scale continuum mechanics methods are developed as an efficient way of modeling CNT. These modeling techniques are not computationally intensive like atomistic modeling and thus they are able to be applied to more complex system with-out limitation of short time and/or length scales. Moreover, the discrete nature of the CNT lattice structure is kept in the modeling by replacing C–C bonds with a continuum element. Since the continuum modeling is employed at the scale of nano, therefore the modeling is called as nanoscale continuum modeling.

The performed investigations in literature addressing mechanical properties, buckling, vibrations and thermal behavior of CNT are reviewed and classified on the basis of three afore mentioned modeling techniques. While atomistic modeling is a reasonable modeling technique for this purpose, its applicability is limited to the small systems. On the other hand, the continuum modeling neglects the discrete structure of CNT leading to inaccurate results.

Nano-scale continuum modeling can be considered as an accept-able compromise in the modeling presenting results in a close agreement with than that of atomistic modeling. Employing FEM as a computationally powerful tool in nanoscale continuum modeling, the influence of CNT chirality, diameter, thickness and other involved parameters can be evaluated conveniently in comparison with other methods. Concerning CNTs buckling properties, many researchers have conducted the simulation using MD methods. The shell theories and molecular mechanic structure simulation are also applied to assess the CNTs buckling in order to avoid time consuming simulations. But, for the specific case of buckling behavior, it is not recommended to scarify the lattice structure of CNT for less time consuming computations. But instead, nanoscale continuum modeling is preferred.

Comparing the results with the obtained results of MD simulation, it can be inferred from literature that for large aspect ratios (i.e., length to diameter ratio $L/d > 10$) the simple Euler–Bernoulli beam is reliable to predict the buckling strains of CNTs while the refined Timoshenko's beam model or their nonlocal counter parts theory is needed for CNTs with intermediate aspect ratios (i.e., $8 < L/d < 10$). The Donnell thin shell theory is incapable to capture the length dependent critical strains for CNTs with small aspect ratios (i.e., $L/d < 8$). On the other hand, Sanders shell theory is accurate in predicting buckling strains and mode shapes of axially compressed CNTs with small aspect ratios. From the dynamic analysis point of view, the replacement of CNT with a hollow cylinder has to be extremely avoided, despite the widely employed method. In other word, replacing CNT with a hollow cylinder will not only lead us to inaccurate results, but also there will not be any difference between and nano-structure in the form of tube with continuum level of modeling. It is a great importance to keep the lattice structure of the CNT in the modeling, since the discrete structure of CNT play an important role in the dynamic analysis. From the vibration point of view, MD simulations are more reliable and NCM approaches are preferred to CM techniques using beam elements.

KEYWORDS

- **Carbon nanotubes**
- **Computational chemistry**
- **Computational mechanics**

REFERENCES

1. Liew, K. M., He, X. Q., & Wong, C. W. (2004). On the Study of Elastic and Plastic Properties of Multi-walled Carbon Nano tubes under Axial Tension using Molecular Dynamics Simulation, *Acta Mater*, *52*, 2521–2527.
2. Zhang, H. W., Wang, J. B., & Guo, X. (2005). Predicting the Elastic Properties of Single-Walled Carbon Nano Tubes, *J Mech Phys Solids*, *53*, 1929–1950.
3. Agrawal Paras, M., Sudalayandi Bala, S., Raff Lionel, M., & Komanduri Ranga (2006). A Comparison of Different Methods of Young's Modulus Determination for Single-Wall Carbon Nanotubes (SWCNT) using Molecular Dynamics (MD) Simulations, *Computer Mater Science*, *38*, 271–281.
4. Cheng Hsien-Chie, Liu Yang Lun, Hsu Yu-Chen, & Chen Wen-Hwa (2009). Atomistic-Continuum Modeling for Mechanical Properties of Single-Walled Carbon Nanotubes, *Int J Solids Struct.*, *46*, 1695–1704.
5. Cai, J., Wang, Y. D., & Wang, C. Y. (2009). Effect of Ending Surface on Energy and Young's Modulus of Single-Walled Carbon Nanotubes Studied using Linear Scaling Quantum Mechanical Method, *Physical B*, *404*, 3930–3934.
6. Ranjbartoreh Ali Reza & Wang Guoxiu (2010). Molecular Dynamic Investigation of Mechanical Properties of Armchair and Zigzag Double-Walled Carbon Nanotubes under Various Loading Conditions, *Phys Lett A*, *374*, 969–974.
7. Sears, A., & Batra, R. C. (2004). Macroscopic Properties of Carbon Nanotubes from Molecular-Mechanics Simulations *Phys Rev B*, *69*, 235406
8. Wang, Q. (2004). Effective in Plane Stiffness and Bending Rigidity of Arm Chair and Zigzag Carbon Nano Tubes, *Int J Solids Struct*, *41*, 5451–5461
9. Kalamkarov, A. L., Georgiades, A. V., Rokkam, S. K., Veedu, V. P., & Ghasemi Nejhad, M. N. (2006). Analytical and Numerical Techniques to Predict Carbon Nanotubes Properties, *Int J Solids Struct.*, *43*, 6832–6854.
10. Gupta, S. S., & Batra, R. C. (2008). Continuum Structures Equivalent in Normal Mode Vibrations to Single-Walled Carbon Nanotubes, *Computer Mater Sci.*, *43*, 715–723.
11. Giannopoulos, G. I., Kakavas, P. A., & Anifantis, N. K. (2008). Evaluation of the Effective Mechanical Properties of Single-Walled Carbon Nanotubes using a Spring Based Finite Element Approach, *Computer Mater Science*, *41(4)*, 561–569.
12. Papanikos, P., Nikolopoulos, D. D., & Tserpes, K. I. (2008). Equivalent Beams for Carbon Nanotubes, *Comput Mater Sci*, *43*, 345–352.
13. Li Chunyu, & Chou Tsu-Wei (2003). Elastic Moduli of Multi-Walled Carbon Nanotubes and the Effect of Van Der Waals Forces, *Compos Sci Technol*, *63*, 1517–15124.
14. Natsuki T., Tantrakarn, K., & Endo, M. (2004) Prediction of Elastic Properties for Single-Walled Carbon Nanotubes, *Carbon*, *42*, 39–45.
15. Natsuki Toshiaki, & Endo Morinobu (2004). Stress Simulation of Carbon Nanotubes Intension and Compression, *Carbon*, *42*, 2147–2151.

16. Xiao, J. R., Gama, B. A., & Gillespie, Jr J. W. (2005). An Analytical Molecular Structural Mechanics Model for the Mechanical Properties of Carbon Nanotubes, *Int J Solids Struct., 42*, 3075–3092.

17. Tserpes, K. I., Papanikos, P. (2005). Finite Element Modeling of Single-Walled Carbon Nanotubes, *Composites Part B, 36*, 468–477.

18. Jalalahmadi, B., & Naghdabadi, R. (2007). Finite Element Modeling of Single-Walled Carbon Nanotubes with Introducing a New Wall Thickness, *J Phys Conf Ser, 61*, 497–502.

19. Meo, M., & Rossi, M. (2006). Prediction of Young's Modulus of Single Wall Carbon Nanotubes by Molecular Mechanics Based Finite Element Modeling, *Compos Sci Technol, 66*, 1597–1605.

20. PourAkbar Saffar Kaveh, JamilPour Nima, Najafi Ahmad Raeisi, RouhiGholamreza, Arshi Ahmad Reza, Fereidoon Abdolhossein, et al. (2008). A Finite Element Model for Estimating Young's Modulus of Carbon Nano Tube Reinforced Composites Incorporating Elastic Cross-Links, *World Acad Sci. Eng Technol, 47*.

21. Cheng Hsien-Chie, Liu Yang-Lun, Hsu Yu-Chen, & Chen Wen-Hwa (2009). Atomistic-Continuum Modeling for Mechanical Properties of Single-Walled Carbon Nanotubes, *Int J Solids Struct, 46*, 1695–1704.

22. Ávila Antonio Ferreira, & Lacerda Guilherme Silveira Rachid (2008). Molecular Mechanics Applied to Single-Walled Carbon Nanotubes, *Mater Res., 11(3)*, 325–333.

23. Wernik Jacob M., & Meguid Shaker, A. (2010). Atomistic-Based Continuum Modeling of the Nonlinear Behavior of Carbon Nanotubes, *Acta Mech, 212*, 167–179

24. Shokrieh Mahmood, M., & Rafiee Roham (2010). Prediction of Young's Modulus of Graphene Sheets and Carbon Nanotubes using Nanoscale Continuum Mechanics Approach, *Mater Des, 31*, 790–795.

25. Lu Xiaoxing, & Hu Zhong (2012). Mechanical Property Evaluation of Single-Walled Carbon Nanotubes by Finite Element Modeling, *Composites Part B, 43*, 1902–1913.

26. Rafiee Roham & Heidarhaei Meghdad (2012). Investigation of Chirality and Diameter Effects on the Young's Modulus of Carbon Nanotubes using Non-Linear Potentials, *Compos Struct, 94*, 2460–2464.

27. Zhang, Y. Y., Tan, V. B. C., & Wang, C. M. (2006). Effect of Chirality on Buckling Behavior of Single-Walled Carbon Nanotubes, *J Appl Phys, 100*, 074304.

28. Chang, T. (2007). Torsional Behavior of Chiral Single-Walled Carbon Nanotubes is Loading Direction Dependent, *Appl Phys Lett, 90*, 201910.

29. Chang Tienchong, Li Guoqiang, & Guo Xingming (2005). Elastic Axial Buckling of Carbon Nanotubes via a Molecular Mechanics Model, *Carbon, 43*, 287–294.

30. Wang, C. M., Zhang, Y. Y., Xiang, Y., & Reddy, J. N. (2010). Recent Studies on Buckling of Carbon Nano Tubes, *Appl Mech Rev., 63*, 030804.

31. Srivastava, D., Menon, M., & Cho, K. J. (1999). Nano Plasticity of Single-Wall Carbon Nanotubes under Uniaxial Compression, *Phys Rev Lett., 83(15)*, 2973–2976.

32. Ni, B., Sinnott, S. B., Mikulski, P. T. et al. (2002). Compression of Carbon Nanotubes Filled with C-60, CH4, or Ne, Predictions from Molecular Dynamics Simulations, Phys Rev Lett, *88*, 205505.

33. Wang, Yu, Wang Xiu-xi, Ni Xiang-gui, & Wu Heng-an (2005). Simulation of the Elastic Response and the Buckling Modes of Single-Walled Carbon Nanotubes, *Comput Mater Sci., 32*, 141–146.

34. Hao Xin, Qiang Han, & Xiaohu Yao (2008). Buckling of Defective Single-Walled and Double-Walled Carbon Nanotubes under Axial Compression by Molecular Dynamics Simulation, *Compos Sci Technol, 68*, 1809–1814.

35. Ansari, R., Sahmani, S., & Rouhi, H. (2011). Rayleigh–Ritz Axial Buckling Analysis of Single-Walled Carbon Nanotubes with Different Boundary Conditions, *Phys Lett A.*, *375*, 1255–1263.
36. Yakobson, B. I., Brabec, C. J., & Bernholc, J. (1996); Nano Mechanics of Carbon Tubes Instabilities beyond Linear Response, *Phys Rev Lett.*, *76*, 2511–2514.
37. Zhang Chen-Li & Shen Hui-Shen (2006). Buckling and Post Buckling Analysis of Single-Walled Carbon Nanotubes in Thermal Environments via Molecular Dynamics Simulation, *Carbon*, *44*, 2608–2616.
38. Jeong Byeong-Woo & Sinnott Susan, B. (2010). Unique Buckling Responses of Multi-Walled Carbon Nanotubes Incorporated as Torsion Springs, *Carbon*, *48*, 1697–1701.
39. Silvestre, N., Wang, C. M., Zhang, Y. Y., & Xiang, Y. (2011). Sanders Shell Model for Buckling of single-Walled Carbon Nanotubes with Small Aspect Ratio, Compos Struct., *93*, 1683–1691.
40. Donnell, L. H. (1933). Stability of Thin-Walled Tubes under Torsion, NACA Report No. 479.
41. Sanders, J. L. (1963). Non-Linear Theories for Thin Shells, *Quart J Appl Math*, *21*, 21–36.
42. Silvestre Nuno, Faria Bruno & Canongia Lopes José, N. (2012). A Molecular Dynamics Study on the Thickness and Post-Critical Strength of Carbon Nanotubes, *Compos Struct*, *94*, 1352–1358.
43. Ghavamian Ali, & Öchsner Andreas (2012). Numerical Investigation on the Influence of Defects on the Buckling Behavior of Single-and Multi-walled Carbon Nanotubes, *Physica E*, *46*, 241–249.
44. Zhang, Y. Y., Wang, C. M., Duan, W. H., Xiang, Y., & Zong, Z. (2009). Assessment of Continuum Mechanics Models in Predicting Buckling Strains of Single-Walled Carbon Nanotubes, *Nanotechnology*, *20*, 395707.
45. Sears, A., & Batra, R. C. (2006). Buckling of Multi Walled Carbon Nanotubes under Axial Compression, *Phys Rev B*, *73*, 085410.
46. Xin Hao, Han Qiang, & Yao Xiao-Hu (2007). Buckling and Axially Compressive Properties of Perfect and Defective Single-Walled Carbon Nanotubes, *Carbon*, *45*, 2486–2495.
47. Silvestre Nuno, Faria Bruno & Canongia Lopes José, N. (2012). A Molecular Dynamics Study on the Thickness and Post-Critical Strength of Carbon Nanotubes, *Compos Struct*, *94*, 1352–1358.
48. He, X. Q., Kitipornchai, S., & Liew, K. M. (2005). Buckling Analysis of Multi-Walled Carbon Nanotubes, a Continuum Model Accounting for Van Der Waals Interaction, *J Mech Phys Solid*, *53*, 303–326.
49. Ghorbanpour Arani, A., Rahmani, R., & Arefmanesh, A. (2008). Elastic Buckling Analysis of Single-Walled Carbon Nano Tube under Combined Loading by Using the ANSYS Software, *Physica E*, *40*, 2390–2395.
50. Yao Xiao hu, Han Qiang, & Xin Hao (2008). Bending Buckling Behaviors of Single and Multi-Walled Carbon Nanotubes, *Comput Mater Sci*, *43*, 579–590.
51. Guo, X., Leung, A. Y. T., He, X. Q., Jiang, H., & Huang, Y. (2008). Bending Buckling of Single-Walled Carbon Nanotubes by Atomic-Scale Finite Element, *Composites Part B.*, *39*, 202–208.
52. Chan Yue, Thamwattana Ngamta, & Hill James, M. (2011). Axial Buckling of Multi-Walled Carbon Nanotubes and Nano Peapods, *Eur J Mech A/Solid*, *30*, 794–806.
53. Silvestre Nuno (2012). On the Accuracy of Shell Models for Torsional Buckling of Carbon Nanotubes, *Eur J Mech A/Solid*, *32*, 103–108.
54. Li Chunyu & Chou Tsu-Wei (2004). Modeling of Elastic Buckling of Carbon Nano Tubes by Molecular Structural Mechanics Approach, *Mech Mater*, *36*, 1047–1055.

55. Hu N., Nunoya, K., Pan, D., Okabe, T., & Fukunaga, H. (2007). Prediction of Buckling Characteristics of Carbon Nanotubes, *Int J Solids Struct, 44,* 6535–6550
56. Sawano, S., Arie, T., & Akita, S. (2010). Carbon Nano Tube Resonator in Liquid, *Nano Lett, 10,* 3395–3398.
57. Formica Giovanni, Lacarbonara Walter, & Alessi Roberto (2010). Vibrations of Carbon Nano Tube-Reinforced Composites, *J Sound Vib, 329,* 1875–1889.
58. Khan Shafi Ullah, Li Chi Yin, Siddiqui Naveed, A., & Kim Jang-Kyo (2011). Vibration Damping Characteristics of Carbon Fiber-Reinforced Composites Containing Multi-Walled Carbon Nanotubes, *Compos Sci Technol, 71,* 1486–1494
59. Gibson, R. F., Ayorinde, E. O., & Wen, Y. F. (2007). Vibrations of Carbon Nanotubes and their Composites, *A Review Compos Sci Technol, 67,* 1–28.
60. Wang, C. M., Tan, V. B. C., & Zhang, Y. Y. (2006). Timoshenko Beam Model for Vibration Analysis of Multi-Walled Carbon Nanotubes, *J Sound Vib, 294,* 1060–1072.
61. Hu Yan-Gao, Liew, K. M., & Wang, Q. (2012). Modeling of Vibrations of Carbon Nano Tubes, *Proc Eng, 31(2012),* 343–347.
62. Li, C., & Chou, T. W. (2003). Single-Walled Carbon Nanotubes as Ultra High Frequency Nano Mechanical Resonators, *Phys Rev B, 68,* 073405
63. Arghavan, S., & Singh, A. V. (2011). On the Vibrations of Single-Walled Carbon Nanotubes, *J Sound Vib., 330,* 3102–3122.
64. Sakhaee-Pour, A., Ahmadian, M. T., & Vafai, A. (2009).Vibrational Analysis of Single-Walled Carbon Nanotubes Using Beam Element, *Thin Wall Struct, 47,* 646–652.
65. Gupta, S. S., Bosco, F. G., & Batra, R. C. (2010). Wall Thickness and Elastic Moduli of Single-Walled Carbon Nanotubes from Frequencies of Axial, Torsional and in Extensional Modes of Vibration, *Comput Mater Sci., 47,* 1049–1059.
66. Aydogdu Metin (2012). Axial Vibration Analysis of Nano Rods (Carbon Nanotubes) Embedded in an Elastic Medium Using Nonlocal Elasticity, *Mech Res Commun, 43,* 34–40.
67. Fakhrabadi Mir Masoud Seyyed, Amini Ali, & Rastgoo Abbas (2012). Vibrational Properties of Two and Three Junctioned Carbon Nanotubes, *Comput Mater Sci., 65,* 411–425.
68. Lee, J. H., & Lee, B. S. (2012). Modal Analysis of Carbon Nanotubes and Nano Cones using FEM, *Computer Mater Sci., 51,* 30–42.
69. Fu, Y. M., Hong, J. W., & Wang, X. Q. (2006). Analysis of Nonlinear Vibration for Embedded Carbon Nanotubes, *J Sound Vib, 296,* 746–756.
70. Ansari, R., & Hemmatnezhad, M. (2011). Non Linear Vibrations of Embedded Multi-walled Carbon Nano Tubes using a Variational Approach, *Math Comput Model, 153,* 927–938.
71. Wang, C.Y., Li C. F., & Adhikari, S. (2010). Axisymmetric Vibration of Single-Walled Carbon Nanotubes in Water, *Phys Lett A., 374,* 2467–2474.
72. Natsuki Toshiaki, Qing-Qing, Ni., & Endo Morinobu (2008). Analysis of the Vibration Characteristics of Double-Walled Carbon Nanotubes, *Carbon, 46,* 1570–1573.
73. Aydogdu Metin (2008). Vibration of Multi-Walled Carbon Nanotubes by Generalized Shear Deformation Theory, *Int J Mech Sci, 50,* 837–844.
74. Lei Xiao-Wen, Natsuki Toshiaki, Shi Jin-Xing, & Qing-Qing, Ni (2011). Radial Breathing Vibration of Double-Walled Carbon Nanotubes Subjected to Pressure, *Phys Lett A, 375,* 2416–2421.
75. Ambrosini Daniel, & de Borbón Fernanda (2012). On the Influence of the Shear Deformation and Boundary Conditions on the Transverse Vibration of Multi-Walled Carbon Nano tubes, *Comput Mater Sci, 53,* 214–219.
76. de Borbón Fernanda, & Ambrosini Daniel (2012). On the Influence of Van der Waals Coefficient on the transverse Vibration of Double Walled Carbon Nano Tubes, *Comput Mater Sci, 65,* 504–508.

77. Gupta, S. S., & Batra, R. C. (2008). Continuum Structures Equivalent in Normal Mode Vibrations to Single-Walled Carbon Nanotubes, *Comput Mater Sci, 43,* 715–723.
78. Ansari, R., Gholami, R., & Rouhi, H. (2012). Vibration Analysis of Single-Walled Carbon Nanotubes using Different Gradient Elasticity Theories, Composites Part B, *43(8),* 2985–2989.
79. Ansari, R., Ajori, S., & Arash, B. (2012).Vibrations of Single and Double-Walled Carbon Nanotubes with Layer-Wise Boundary Conditions, a Molecular Dynamics Study, *Curr Appl Phys 12,* 707–711.
80. Sun, C., & Liu, K. (2007). Vibration of Multi-Walled Carbon Nanotubes with Initial Axial Loading, *Solid State Commun, 143,* 202–207.
81. Ke, L. L., Xiang, Y., Yang, J., & Kitipornchai, S. (2009). Nonlinear Free Vibration of Embedded Double-Walled Carbon Nanotubes Based on Nonlocal Timoshenko Beam Theory, *Comput Mater Sci., 47,* 409–417.
82. Ghavanloo, E., & Fazelzadeh, S. A. (2012). Vibration Characteristics of Single-Walled Carbon Nanotubes Based on an Anisotropic Elastic Shell Model Including Chirality Effect, *Appl Math Model, 36,* 4988–5000.
83. Khosrozadeh, A., & Hajabasi, M. A. (2012). Free Vibration of Embedded Double-Walled Carbon Nanotubes Considering Nonlinear Interlayer Van Der Waals Forces, *Appl Math Model, 36,* 997–1007.
84. Georgantzinos, S. K., Giannopoulos, G. I., & Anifantis, N. K. (2009). An Efficient Numerical Model for Vibration Analysis of Single-Walled Carbon Nanotubes, *Comput Mech, 43,* 731–741.
85. Georgantzinos, S. K., & Anifantis, N. K. (2009). Vibration Analysis of Multi-Walled Carbon Nanotubes using a Spring-Mass Based Finite Element Model, *Computer Mater Sci, 47,* 168–77.

CHAPTER 14

CONTROLLED IMMOBILIZED ENZYMES AS CATALYSTS WITH PARTICULAR APPLICATION IN INDUSTRIAL CHEMICAL PROCESSES

M. S. MOHY ELDIN, M. R. EL-AASSAR, and E. A. HASSAN

CONTENTS

ABSTRACT

In recent years, enzyme immobilization has gained importance for design of artificial organs, drug delivery systems, and several biosensors. Polysaccharide based natural biopolymers used in enzyme or cell immobilization represent a major class of biomaterials which includes agarose, alginate, dextran, and chitosan. Especially, Alginates are commercially available as water-soluble sodium alginates and they have been used for more than 65years in the food and pharmaceutical industries as thickening, emulsifying and film forming agent. Entrapment within insoluble calcium alginate gel is recognized as a rapid, nontoxic, inexpensive and versatile method for immobilization of enzymes as well as cells. In this research, the formulation conditions of the alginate beads entrapment immobilized with the enzyme have been optimized and effect of some selected conditions on the kinetic parameter, Km, have been presented. β-galactosidase enzymes entrapped into alginate beads are used in the study of the effect of both substrate diffusion limitation and the mis-orientation of the enzyme on its activity, the orientation of an immobilized protein is important for its function. Physicochemical characteristics and kinetic parameters; Protection of the activity site using galactose as protecting agent has been presented as a solution for the mis-orientation problem. This technique has been successful in reduction, and orientation-controlled immobilization of enzyme. Other technique has been presented to reduce the effect of substrate diffusion limitation through covalent immobilization of the enzyme onto the surface of alginate beads after activation of its OH-groups.

14.1   INTRODUCTION

Recently an increasing trend has been observed in the use of immobilized enzymes as catalysts in several industrial chemical processes. Immobilization is important to maintain constant environmental conditions order to protect the enzyme against changes in pH, temperature, or ionic strength; this is generally reflected in enhanced stability [1]. Moreover, immobilized enzymes can be more easily separated from substrates and reaction products and used repeatedly. Many different procedures have been developed for enzyme immobilization; these include adsorption to insoluble materials, entrapment in polymeric gels, encapsulation in membranes, cross linking with a bi functional reagent, or covalent linking to an insoluble carrier [2]. Among these, entrapment in calcium alginate gel is one of the simplest methods of immobilization. The success of the calcium alginate gel entrapment technique is due mainly to the gentle environment it provides for the entrapped material.

However, there are some limitation such as low stability and high porosity of the gel [1]. These characteristics could lead to leakage of large molecules like proteins, thus generally limiting its use to whole cells or cell organelles [3].

Enzymes are biological catalysts with very good prospects for application in chemical industries due to their high activity under mild conditions, high selectivity and specificity [4–7]. However, enzymes do not fulfill all of the requirements of an industrial biocatalyst or biosensor [8, 9]. They have been selected throughout natural evolution to perform their physiological functions under stress conditions and quite strict regulation. However, in industry, these biocatalysts should be heterogeneous, reasonably stable under conditions that may be quite far from their physiological environment [10] and retain their good activity and selectivity when acting with substrates that are in some instances quite different from their physiological ones.b-galactosidase, commonly known as lactase, catalyzes the hydrolysis of b-galactosidic linkages such as those between galactose and glucose moieties in the lactose molecule. While the enzyme has many analytical uses, being a favorite label in various affinity recognition techniques such as ELISAs, or enzyme-linked immunosorbent assays, its main use is the large scale processing of dairy products, whey, and whey permeates.

Immobilization of the enzymes to solid surface induces structural changes which may affect the entire molecule. The study of conformational behavior of enzymes on solid surface is necessary for better understanding of the immobilization mechanism. However, the immobilization of enzymes on alginate beads is generally rapid, and depends on hydrophobic and electrostatic interactions as well as on external conditions such as pH, temperature, ionic strength and nature of buffer [11, 12]. Enzymes denaturation may occur under the influence of hydrophobic interactions, physicochemical properties of the alginate beads or due to the intrinsic properties of the enzyme.

One of the main concerns regarding immobilized proteins has been the reduction in the biological activity due to immobilization. The loss in activity could be due to the immobilization procedure employed, changes in the protein conformation after immobilization, structural modification of the protein during immobilization, or changes in the protein microenvironment resulting from the interaction between the support and the protein. In order to retain a maximum level of biological activity for the immobilized protein, the origins of these effects need to be understood, especially as they relate to the structural organization of the protein on the immobilization surface. The immobilization of proteins on surfaces can be accomplished both by physical and chemical methods. Physical methods of immobilization include the attachment of protein to surfaces by various inter actions such as electrostatic, hydrophobic/hydrophilic and Van der Waals forces. Though the method is simple and cost-effective, it suffers from protein leaching from the immobilization support. Physical adsorption generally leads to dramatic changes in the protein microenvironment, and typically involves multipoint protein adsorption between a single protein molecule and a number of binding sites on the immobilization surface. In addition, even if the surface has a uniform distribution of binding sites, physical adsorption could lead to heterogeneously populated immobilized proteins. This has been ascribed to unfavorable lateral interactions among bound protein molecules [13]. The effect of ionic strength on protein adsorption has also been studied, and a striking dependence on the concentration of electrolyte was noticed. For example, in the case of adsorption of a transfer in on a silicon titanium dioxide surface, it was found that an increase in ionic strength resulted in a decrease in adsorption. The increase in ionic strength decreases the negative surface potential and increases the surface pH. This leads to an increase in the net protein charge, creating an increasingly repulsive energy barrier, which results in reduced effective diffusivity of molecules to the surface [14]. Several research groups have attempted to develop a model that can qualitatively and quantitatively predict physical adsorption, especially as it relates to ion-exchange chromatography. Models have been developed to predict average interaction energies and preferred protein orientations for adsorption. Adsorption studies on egg white lysozyme and o-lactalbumin adsorbed on anion-exchanger polymeric surfaces have shown that it is possible to determine the residues involved in the interaction with the support [15, 16]. It has been also possible to calculate the effective net charge of the proteins. Entrapment and microencapsulation are other popular physical methods of immobilization and have been discussed elsewhere [17]. The immobilization of enzymes through metal chelation has been reviewed recently [18].

Attachment of proteins by chemical means involves the formation of strong covalent or coordination bonds between the protein and the immobilization support. The chemical attachment involves more drastic (non mild) conditions for the immobilization reaction than the attachment through adsorption. This can lead to a significant loss in enzyme activity or in binding ability (in the case of immobilization of binding proteins and antibodies). In addition, the covalent and coordinate bonds formed between the protein and the support can lead to a change in the structural configuration of the immobilization protein. Such a change in the enzymatic structure may lead to reduced activity, unavailability of the active site of an enzyme for the substrates, altered reaction pathways or a shift in optimum pH [19, 20]. In the case of immobilized antibodies and binding proteins this structural change can lead to reduced binding ability.

Oriented immobilization is one such approach [21]. Adsorption, bioaffinity immobilization and entrapment generally give high retention of activity as compared to covalent coupling method [22–26]. Use of macro porous matrices also helps by reducing mass transfer constraints. Calcium alginate beads, used here, in that respect, are an attractive choice. Calcium alginate beads are not used for enzyme immobilization as enzymes slowly diffuse out [31]. In this case, binding of b-galactosidase and galactos to alginate before formation of calcium alginate beads ensured the entrapment of these individual enzymes and undoubtedly, also contributed to enhanced thermo stabilization. Such a role for alginate in fact, has already been shown for another alginate binding enzyme.

In this work, b-galactosidase enzyme entrapped into alginate beads are used in the study of the effect of both substrate diffusion limitation and the mis-orientation of the enzyme on its activity, Physicochemical characteristics and kinetic parameters. Protection of the activity site using galactose as protecting agent has been presented as a solution for the mis-orientation problem. This technique has been successful in reduction, but not in elimination of the effect of mis-orientation on the activity as well as the kinetic parameters; Km and Vm. Other technique has been presented to reduce the effect of substrate diffusion limitation through covalent immobilization of the enzyme onto the surface of alginate beads after activation of its OH-groups. The impact of different factors controlling the activation process of the alginate hydroxyl groups using p-benzoquinone (PBQ) in addition to the immobilization conditions on the activity of immobilized enzyme have been studied. The immobilized enzyme has been characterized from the bio-chemical point of view as compared with the free enzyme.

14.2 MATERIALS AND METHODS

14.2.1 MATERIALS

Sodium alginate (low viscosity 200cP), β-Galactosidase (from Aspergillus oryzae) p-Benzoquinone (purity 99+%):Sigma-Aldrich chemicals Ltd. (Germany); calcium chloride(anhydrous Fine GRG90%): Fisher Scientific (Fairlawn, NJ, USA); ethyl alcohol absolute, Lactose (pure Lab. Chemicals, MW 360.31), Galactose: El-Nasr Pharmaceutical Chemicals Co. (Egypt); glucose kit (Enzymatic colorimetric method):Diamond Diagnostics Co. for Modern Laboratory Chemicals(Egypt); Tris-Hydrochloride (Ultra Pure Grade 99.5%,

MW 157.64): amresco (Germany); sodium chloride (Purity 99.5%): BDH Laboratory Supplies Pool (England).

14.2.2   METHODS

14.2.2.1   PREPARATION OF CATALYTIC CA-ALGINATE GEL BEADS BY ENTRAPMENT TECHNIQUE

The Ca- alginate gel beads prepared by dissolving certain amount of sodium alginate (low viscosity) 2%w/v in distilled water with continuous heating the alginate solution until it becomes completely clear solution, mixing the alginate solution with enzyme [β-galactosidase enzyme as a model of immobilized enzyme (0.005 g)]. The sodium alginate containing β-galactosidase is dropped (drop wise) by 10 cm^3 plastic syringe in (50 mL) calcium chloride solution (3%w/v) as a safety cross linker to form beads to give a known measurable diameter of beads. Different aging times of beads in calcium chloride solution are considered followed by washing the beads by (50 mL) buffer solution, and determination of the activity of immobilized enzyme. In case of oriented immobilization, galactose with different concentrations was mixed first with the enzyme-buffer solution before mixing with the alginate.

14.2.2.2   ALGINATE BEADS SURFACE MODIFICATION

The Ca-alginate gel beads prepared by dissolving sodium alginate (low viscosity) in distilled water with continuous heating the solution until become completely clear to acquire finally 4% (w/v) concentration. The alginate solution was mixed with PBQ solution (0.02 M) and kept for four hours at room temperature to have final concentration 2% (w/v) alginate and 0.01 M (PBQ). The mixture was added drop wise, using by 10 cm^3 plastic syringes to calcium chloride solution (3%w/v) and left to harden for 30 min at room temperature to reach 2 mm diameter beads. The beads were washed using buffer-ethanol solution (20% ethanol) and distilled water, to remove the excess (PBQ), before transferring to the enzyme solution (0.005 g of β-galactosidase in 20 mL of Tris-HCl buffer solution of pH=4.8) and stirring for 1 h at room temperature then the mixture was kept at 4 °C for 16 h to complete the immobilization process. The mechanism of the activation process and enzyme immobilization is presented in Scheme 14.1.

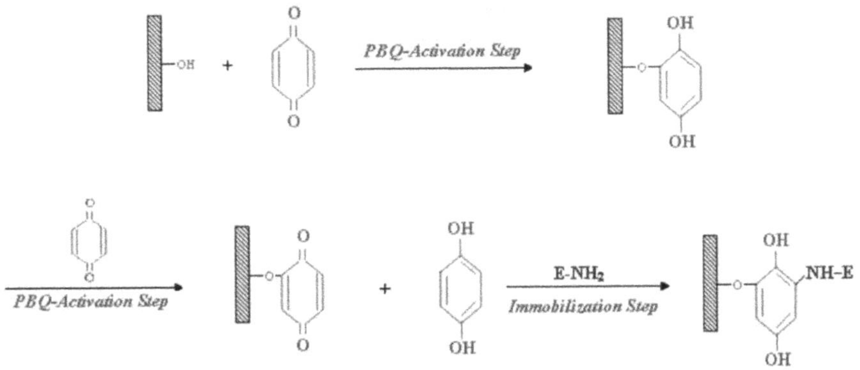

SCHEME 14.1 Mechanism of Activation and Immobilization Process.

14.2.2.3 DETERMINATION OF IMMOBILIZED ENZYME ACTIVITY

The catalytic beads were mixed with 0.1 M Lactose-Tris-HCl buffer solution of (pH 4.8 with stirring, 250 rpm, at room temperature for 30 min. Samples were taken every 5 min to assess the glucose production using glucose kit. Beads activity is given by the angular coefficient of the linear plot of the glucose production as a function of time.

14.3 RESULTED AND METHODS

14.3.1 ENTRAPMENT IMMOBILIZATION

The impact of different factors affecting the process of enzyme entrapment and its reflection on the activity of the immobilized enzyme, its physicochemical characters and its kinetic parameters have been studied and the obtained results are given below.

14.3.1.1 EFFECT OF ALGINATE CONCENTRATION

It's clear from Fig. 14.1 that increasing the concentration of alginate has a linear positive effect on the activity. Increasing the activity with alginate concentration can be explained in the light of increasing the amount of entrapped

enzyme as a result of the formation of as more densely cross-linked gel structure [28]. This explanation has been confirmed by the data obtained in case of
immobilized and/or entrapped amount of enzyme.

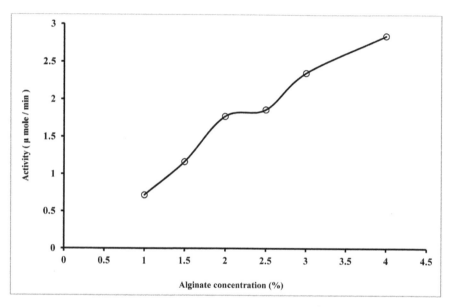

FIGURE 14.1   Effects of alginates concentrations on beads catalytic activity.

14.3.1.2   EFFECT OF CROSS-LINKING TEMPERATURE

Opposite behavior of the activity has been obtained with increasing the temperature of cross-linking process in CaCl$_2$ solution (Fig. 14.2). Such results
could be explained based on increasing cross-linking degree along with the
temperature of calcium chloride solution, which leads finally to decrease the
amount of diffusive lactose substrate to the entrapped enzyme and hence reduce the activity of the immobilized enzyme.

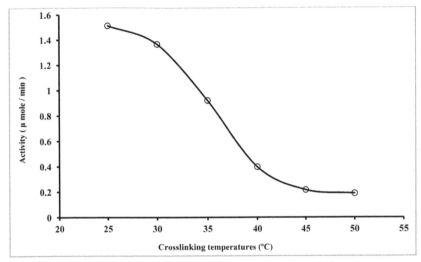

FIGURE 14.2   Effects of cross-linking temperature on beads catalytic activity.

14.3.1.3   EFFECT OF CROSS-LINKING TIME

Figure 14.3 shows the effect of increasing the cross-linking time in $CaCl_2$ solution on the activity of entrapped enzyme. It's clear that a reduction of about 20% of the activity has been detected with increasing the time from 30 to 60 min. Further increase has no noticeable effect on the activity.

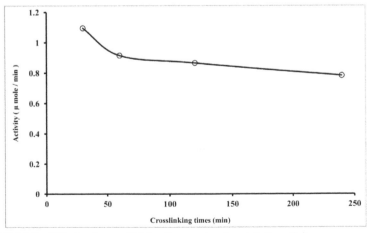

FIGURE 14.3   Effects of cross-linking time on beads catalytic activity.

14.3.1.4   EFFECT OF CACL₂ SOLUTION CONCENTRATION

The effect as given in Fig. 14.4 of $CaCl_2$ concentration on the activity of en-trapped enzyme shows that increase in $CaCl_2$ concentration decreases the ac-tivity linearly. Such results could be interpreted in the light of increasing the gel cross-linking density and hence reducing the amount of diffusive lactose.

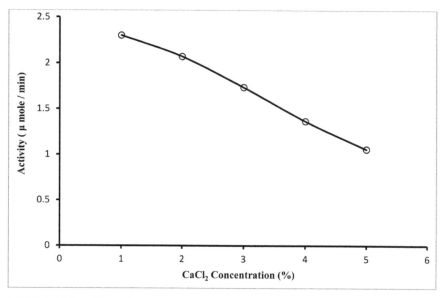

FIGURE 14.4   Effect of $CaCl_2$concentration (%) on beads catalytic activity.

14.3.1.5   EFFECT OF ENZYME CONCENTRATION

The dependence of the catalytic activity of the beads on enzyme concentration is illustrated in Fig. 14.5. It is clear that increasing the enzyme concentration increases the activity linearly within the studied range. Increasing the amount of entrapped enzyme is the logic explanation of such results.

14.3.1.6   EFFECT OF SUBSTRATE'S TEMPERATURE

Shift of the optimum temperature towards lower side has been detected upon entrapping of the enzyme (Fig. 14.6). A suitable temperature found to be

45 °C for the entrapped enzyme compared with 55 °C for the free form. Such behavior may be explained in the light of acidic environment due to the presence of free unbinding carboxylic groups. This explanation is confirmed by the higher rate of enzyme denaturation at higher temperatures, 50–70 °C, in comparison with the free form.

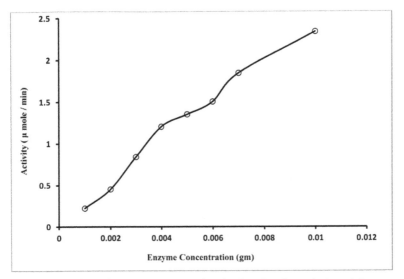

FIGURE 14.5   Effect of enzyme concentration (gm) on beads catalytic activity.

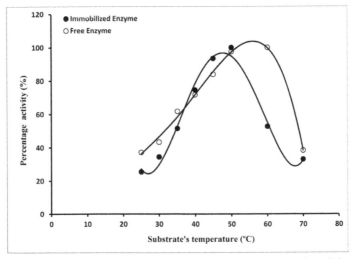

FIGURE 14.6   Effect of substrate's temperature (°C) on beads catalytic activity.

14.3.1.7   EFFECT OF SUBSTRATE'S PH

Similar behavior of the immobilized enzyme to the free one has been observed in respond to the changes in the substrate's pH (Fig. 14.7). The optimum pH has not shifted, but the immobilized enzyme shows higher activity within acidic region, pH 2.5–3.0. This resistance to pH could be explained due to the presence of negatively charged free carboxylic groups un-binding by calcium ions. This result is in accordance with those obtained by other authors [29].

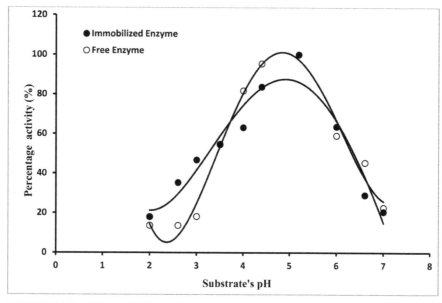

FIGURE 14.7   Effect of Substrate's pH on Beads Catalytic Activity.

14.3.1.8   KINETIC STUDIES

Figure 14.8 presents the effect of the immobilization process on the kinetic parameters especially *Km* which can be considered as the reflection of presence or absence of substrate diffusion limitation.

A Han's plot curve indicates how much the immobilization process can be affected as medicated from the kinetic parameters of immobilized enzyme. That the Km value has become five times higher than that of the free one is a reflection of powerful diffusion limitation of the substrate.

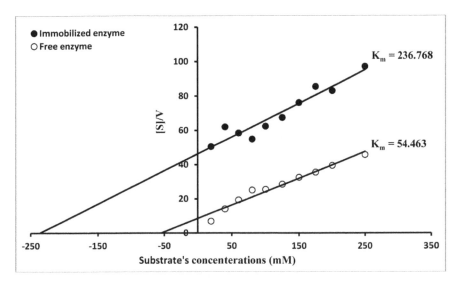

FIGURE 14.8   Han's Plot Curve for Free and Immobilized Enzyme.

In addition, mis-orientation of the active sites as a result of the immobilization process could be another factor affecting the accessibility of the active sites to the substrate. To study this effect, galactose has been added during the immobilization process in different concentrations to protect the active site (Fig. 14.9).

The Km values have been reduced with increasing the galactose concentration reaching minimum value at 30μ mole (Fig. 14.16). Beyond this concentration, the Km value starts to increase again.

The protection of the active sites has a positive reflection on the activity of immobilized enzyme (Fig. 14.10). The reduction in the Km value and activity enhancement proved the role of mis-orientation but still the fact that diffusion limitation of the substrate has the main role.

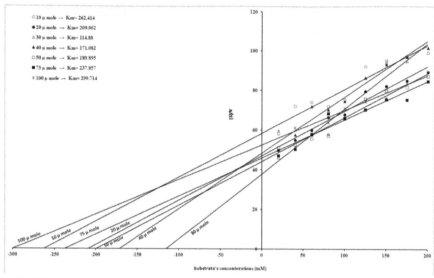

FIGURE 14.9   Han's Plot Curve for Immobilized Enzyme with Addition of Galactose.

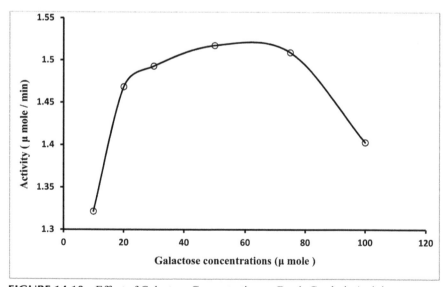

FIGURE 14.10   Effect of Galactose Concentration on Beads Catalytic Activity.

To solve the problem of diffusion limitation, the enzyme was used to be covalently immobilized on the surface of the alginate beads after activation with *p*-benzoquinone. The impact of different factors controlling the activation process of the alginate hydroxyl groups using *p*-benzoquinone (PBQ) in addition to the immobilization conditions on the activity of immobilized enzyme have been studied. The immobilized enzyme has been characterized from the bio-chemical point of view as compared with the free enzyme.

14.3.2 ACTIVITY OF THE BEADS

14.3.2.1. THE EFFECT OF ALGINATE CONCENTRATION

From Fig. 14.11, it's clear that increasing the concentration of alginate has negative effect on the activity. A reasonable explanation could be obtained by following the amount of immobilized enzyme, which decreases in the same manner. This could be explained as a result of increasing the cross linking density and a direct reduction of the available pores surface area for enzyme immobilization. Indeed, the retention of activity has not affected so much since the decrease rate of both activity and amount of immobilized enzyme is almost the same [30].

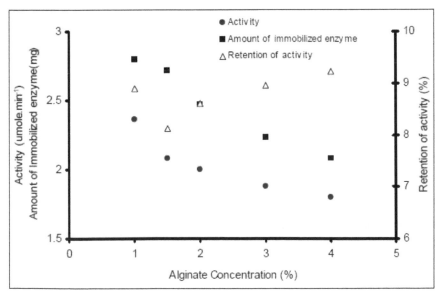

FIGURE 14.11   Effect of alginate concentrations on beads catalytic activity.

14.3.2.2   EFFECT OF CACL$_2$ CONCENTRATION

As shown in Fig. 14.12, the activities of the beads have not been affected so much by changing the concentration of CaCl$_2$ solution. Unexpectedly, the amount of immobilized enzyme has been increased gradually with concentration of CaCl$_2$ solution. This behavior could be explained according to the impact of concentration of CaCl$_2$ solution on the number of the formed pores inside the beads which offering additional surface area for enzyme immobilization. Since the activity was found almost constant and the amount of immobilized enzyme increased under the same conditions, so it is logic to have a reduction of the retention of activity.

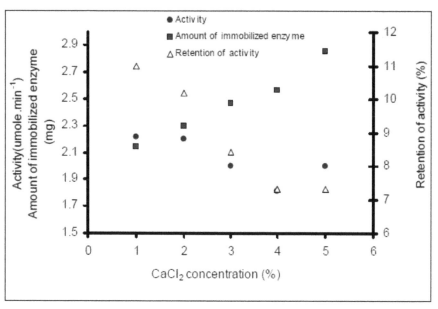

FIGURE 14.12   Effect of CaCl$_2$ concentrations on beads catalytic activity.

14.3.2.3   EFFECT OF AGING TIME

It has been found that changing the aging time from 30 to 240 min has a clear effect on the catalytic activity of the beads. This may be due to the cross linking effect of CaCl$_2$ which affected directly on the pores size producing pores smaller than the enzyme size and hence reducing the available pores surface area for immobilizing enzyme. Since the amount of immobilized enzyme reduced with higher rate than the activity did, so the retention of activity affected positively with aging time increase (Fig. 14.13). It is clear from the results here that not only the pores surface area but also the pores size affects the amount of immobilized enzyme and hence the activity.

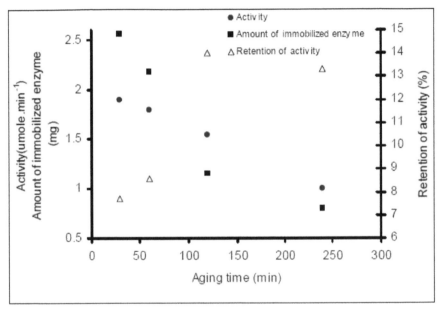

FIGURE 14.13   Effect of aging time on beads catalytic activity.

14.3.2.4   EFFECT OF AGING TEMPERATURE

The dependence of the catalytic activity of the beads on aging temperature is illustrated in Fig. 14.14. From the figure, it is clear that increasing the aging temperature resulted in increasing the activity in gradual way regardless the cross bonding decrease of the amount of immobilized enzyme. This takes us again to the combination between the pores size distribution and pores surface area. The best retention of activity has been obtained with the highest aging temperature.

14.3.2.5   KINETIC STUDIES

The kinetic parameters, K_m and V_m, of the immobilized enzyme under selected conditions, aging time and temperature, have been calculated from the Hanes Plot curves and tabulated in the following tables, Tables 14.1 and 14.2, respectively. In addition, the Diffusion coefficient (De) of lactose substrate has been calculated [31] and related to the K_m and V_m values.

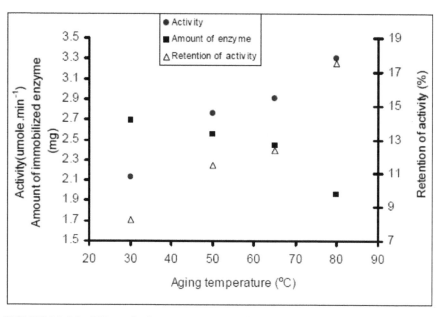

FIGURE 14.14   Effect of aging temperatures on beads catalytic activity.

TABLE 14.1   Kinetic parameters and Diffusion Coefficient of Immobilized Enzyme Prepared Under Different Aging Time

Parameters	Aging time (minutes)			
	30	**60**	**120**	**240**
Km (mmol)	61.00	64.66	133.86	221.80
Vm (umol.min^{-1})	2.94	2.46	4.00	3.36
De (cm^2.s^{-1})	4.9×10^{-11}	4.6×10^{-11}	9.73×10^{-11}	0.184×10^{-11}

TABLE 14.2 Kinetic Parameters and Diffusion Coefficient of Immobilized Enzyme Prepared Under Different Aging Temperature

Parameters	Aging temperature (°C)			
	30	50	65	80
Km (mmol)	61.32	73.44	79.17	265.70
Vm (umol.min^{-1})	2.99	3.24	2.27	5.62
De (cm^2.s^{-1})	5.49×10^{-11}	13.98×10^{-11}	7.3×10^{-11}	10.8×10^{-11}

From the obtained results it's clear that increasing the cross linking degree of the beads through increasing either the aging time or aging temperature in CaCl$_2$ solution has affected directly the Km values which tended to increase. This could be explained based on the diffusion limitation of the lactose into the pores of the beads. This conclusion could be supported by taking into account the obtained data of diffusion coefficient, which in general is higher than the value of immobilized enzyme on beads free of diffusion limitation of the substrate; 1.66×10^{-8} cm^2.s^{-1}. On the other hand V_m values are in general less than that of the free enzyme regardless that some of the obtained K_m values are equal to that of the free enzyme.

14.3.3   BIOCHEMICAL CHARACTERIZATION OF THE CATALYTIC ALGINATE BEADS

14.3.3.1   EFFECT OF SUBSTRATE'S TEMPERATURE

Similar temperature profile to the free enzyme has been obtained for our immobilized form (Fig. 14.15). An optimum temperature has been obtained at 55 °C. This similarity indicates that the microenvironment of the immobilized enzyme is identical to that of the free one, so absence of internal pore diffusion; illustrates the main drawback of the immobilization process, and the success of surface immobilization to overcome this problem. The confirmation of this conclusion has come from determination of the activation energy for both the free and immobilized form [29], which was found very close in value; 6.05 Kcal/mol for the free and 6.12 Kcal/mol for the immobilized forms respectively.

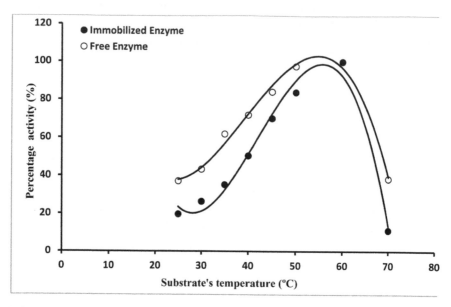

FIGURE 14.15   Effect of substrate's temperature on beads catalytic activity.

14.3.3.2   EFFECT OF SUBSTRATE'S PH

Different behaviors of the immobilized enzyme compared with those of the free one have been observed in respond to the change in the catalytic reaction's pH (Fig. 14.16). Both slight and broaden of the optimum pH for the immobilized form has been obtained since the "optimum activity range," activity from 90–100%, was found from pH value of 4.0 to 5.5. This shift to the acidic range could be explained by the presence of negative charges of unbinding carboxylic groups [29]. This behavior is advantage for use this immobilized form in the degradation of lactose from whey waste, which is normally has a low pH. It's enough to mention here that at (pH 3.0) the immobilized form beads retained 65%, of its maximum activity compared with 20% of the free enzyme.

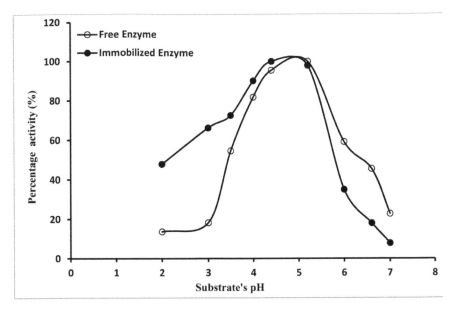

FIGURE 14.16   Effect of substrate pH on beads catalytic activity.

14.3.3.3   OPERATIONAL STABILITY

The operational stability of the catalytic beads was investigated (Fig. 14.17). The beads were repeatedly used for 21 cycles, one hour each, without washing the beads between the cycles. The activity was found almost constant for the first five cycles. After that, gradual decrease of the activity has been noticed. The beads have lost 40% of their original activity after 21 h of net working time. No enzyme was detected in substrate solution, so we cannot claim the enzyme leakage as the cause for activity decline. Incomplete removal of the products from the pores of the beads which can cause masking of the enzyme active sites could be a proper explanation. Washing the beads with suitable buffer solution between the cycles and using higher stirring rate may help in solving this problem.

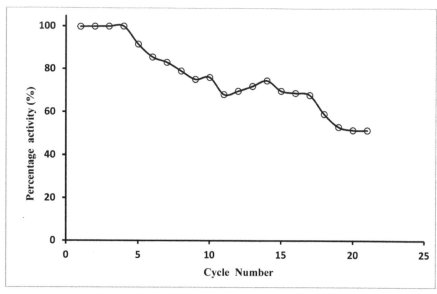

FIGURE 14.17   Operational stability of the catalytic beads.

From the obtained results we can conclude that the activation of alginate's OH groups by using p-benzoquinone offers a new matrix for covalent immobilization of β-galactosidase enzyme. Both the temperature and the pH of the activation process were found to be determining factors in obtaining high catalytic activity beads. The pH of the enzyme solution is found to have a pronounced effect on the catalytic activity of the beads. Three hours of immobilization time was found optimum using 5 mg of enzyme-buffer solution (pH range 4.0–4.5). The wide "optimum pH range" of the activity of the immobilized β-galactosidase makes it's to be recommended to be used in the degradation of lactose in whey waste to reduce its BOD and the production of de-lactose milk to serve the peoples suffering from lactose tolerance. Finally, the immobilization of β-galactosidase on the surface of p-benzoquinone activated alginate beads has succeeded in avoiding the diffusion limitation of lactose.

KEYWORDS

- Bio catalysis
- Catalytic activity beads
- Degradation
- Enzyme immobilization
- Industrial Chemistry
- Pharmaceutical Processes

REFERENCES

1. Cheetham, P. S. J. (1985). Principles of Industrial Enzymology, Basis of Utilization of Soluble and Immobilized Enzymes in Industrial Processes, *Handbook of Enzyme Biotechnology* (Wiseman, A., Ed.) Ellis Horwood Limited, Chichester, 74–86.
2. Klibanov, A. (1983). Immobilized Enzymes and Cells as Practical Catalysts, Science, *219*, 722–727.
3. Smidsrod, O., & Skjak Braek G. (1990). Alginate as Immobilization Matrix for Cells, *Trends Biotech, 8,* 71–78.
4. Pollard, D. J., & Woodley, J. M. (2007). Bio catalysis for Pharmaceutical Intermediates, the Future is Now Trends. *Biotechnol, 25,* 66–73.
5. Woodley, J. M. (2008). New Opportunities for Bio catalysis Making Pharmaceutical Processes Greener, Trends *Biotechnol 26,* 321–327.
6. Schmid, A., Dordick, J. S., Hauer, B., Kiener, A., Wubbolts, M., & Witholt, B. (2001). Industrial Bio catalysis today and Tomorrow, *Nature, 409,* 258–268.
7. Straathof, A. J. J., Panke, S., & Schmid, A. (2002). The Production of Fine Chemicals by Bio transformations, *Curr Opin Biotechnol, 13,* 548–556.
8. Schoemaker, H. E., Mink, D., & Wubbolts, M. G. (2003). Dispelling the Myths Biocatalysis in Industrial Synthesis, *Science, 299,* 1694–1697.
9. Meyer, H. P. (2006). Chemo Catalysis and Bio Catalysis (Biotransformation), Some Thoughts of a Chemist and of a Biotechnologist, *Org Proc Res Dev, 10,* 572–580.
10. Durand, J., Teuma, E., & Gómez, M. (2007). Ionic Liquids as a Medium for Enantioselective Catalysis, *C. R Chim, 10,* 152–177.
11. Trivedi, A. H., Spiess, A. C., Daussmann, T., & Buchs, J. (2006). Effect of Additives on Gas-Phase Catalysis with Immobilized Thermo Anaerobacter Species Alcohol Dehydrogenase ADH T, *Appl. Microbiol Biotechnol, 71,* 407–414.
12. Solis, S., Paniagua, J., Martinez, J. C., & Asmoza, M. (2006). Immobilization of Papa in on Mesoporous Silica, pH Effect, *J. Sol-Gel Sci. Technol. 37,* 125–127.
13. Johnson, R. D., Wang, A. G., & Arnold, E. H. (1996). *J. Phys. Chem., 100,* 5134.
14. Ramsden, J. J., & Prenosil, J. E. (1994). *J. Phys. Chem., 98,* 5376.
15. Noinville, V., Calire, V. M., & Bernard, S. (1995). *J. Phys. Chem, 99,* 1516.
16. Sadana, A. (1993). *Bioseparation, 3,* 297.
17. Marek, P. K. J., & Coughlan, M. E. (1990). In Protein Immobilization (R. E. Taylor, ed.), Marcel Dekker, New York, 13–71.
18. Kennedy, J. E., & Cabral, J. M. S. (1995). Art if Cells, Blood Substitutes Immobilization *Biotechnol, 23,* 231.

19. Mosbach, K., & Mattiasson, B. (1976). Methods Enzy Mol, *44*, 453.
20. Trevan, M. D. (1980). Immobilized Enzymes an Introduction and Applications in Biotechnology, Wiley, Chichester.
21. Wilchek, M., & Miron, T. (2003). Oriented Versus Random Protein Immobilization, *J Biochem Biophys Methods, 55*, 67–70.
22. Gupta, M. N., & Mattiasson, B. (1992). Unique Applications of Immobilized Proteins in Bioanalytical Systems, in Suelter, C. H., & Kricka, L. Bioanalytical Applications of Enzymes, *36*, New York: John Wiley & Sons Inc, 1–34.
23. Mohy Eldin, M. S., Serour, E., Nasr, M., & Teama, H. (2011). *J Appl Biochem Biotechnol, 164*, 10.
24. Mohy Eldin, M. S., Serour, E., Nasr, M., & Teama, H. (2011). *J Appl Biochem Biotechnol, 164*, 45.
25. Mohy Eldin, M. S. (2005). Deutsch Lebensmittel-Rundschau, *101*, 193.
26. Mohy Eldin, M. S., Hassan, E. A., & Elaassar, M. R. (2005). Deutsch Lebensmittel-Rundschau, *101*, 255.
27. Smirsod, O., & Skjak Braek, G. (1990). Alginate as Immobilization Matrix for Cells, Trends, *Biotechnol, 8*, 71–78.
28. Maysinger, D., Jalsenjak, I., & Cuello, A. C. (1992). *Neurosci Lett, 140*, 71–74.
29. Bergamasco, R., Bassetti, F. J., De Moraes, F., & Zanin, G. M. (2000). *Braz. J. Chem. Eng., 17*, 4.
30. Mohy Eldin, M. S., Hassan, E. A., & Elaassar, M. R. December (2004). β-Galactosidase Covalent Immobilization on the Surface of P-Benzoquinone-Activated Alginate Beads as a Strategy for Overcoming Diffusion Limitation III, Effect of Beads Formulation Conditions on the Kinetic Parameters, the 1st International Conference of Chemical Industries Research Division, 6–8.
31. White, C. A., & Kennedy, J. F. (1980). *Enzyme Microb Technol. 2*, 82–90.

INDEX